THE WHISPERER IN THE
DARKNESS

H.P. Lovecraft

Darkness

by H. P. Lovecraft

First published in *Weird Tales*, 1931

Chapter 1

Bear in mind closely that I did not see any actual visual horror at the end. To say that a mental shock was the cause of what I inferred--that last straw which sent me racing out of the lonely Akeley farmhouse and through the wild domed hills of Vermont in a commandeered motor at night--is to ignore the plainest facts of my final experience. Notwithstanding the deep things I saw and heard, and the admitted vividness the impression produced on me by these things, I cannot prove even now whether I was right or wrong in my hideous inference. For after all Akeley's disappearance establishes nothing. People found nothing amiss in his house despite the bullet-marks on the outside and inside. It was just as though he had walked out casually for a ramble in the hills and failed to return. There was not even a sign that a guest had been there, or that those horrible cylinders and machines had been stored in the study. That he had mortally feared the crowded green hills and endless trickle of brooks among which he had been born and reared, means nothing at all, either; for thousands are subject to just such morbid fears. Eccentricity, moreover, could easily account for his strange acts and apprehensions toward the last.

The whole matter began, so far as I am concerned, with the historic and unprecedented Vermont floods of November 3, 1927. I was then, as now, an instructor of literature at Miskatonic University in Arkham, Massachusetts, and an enthusiastic amateur student of New England folklore. Shortly after the flood, amidst the varied reports of hardship, suffering, and organized relief which filled the press, there appeared certain odd stories of things found floating in some of the swollen rivers; so that many of my friends embarked on curious discussions and appealed to me to shed what light I could on the subject. I felt flattered at having my folklore study taken so seriously, and did what I could to belittle the wild, vague tales which seemed so clearly an outgrowth of old rustic superstitions. It amused me to find several persons of education who insisted that some stratum of obscure, distorted fact might underlie the rumors.

The tales thus brought to my notice came mostly through newspaper cuttings; though one yarn had an oral source and was repeated to a friend of mine in a letter from his mother in Hardwick, Vermont. The type of thing described was essentially the same in all cases, though there seemed to be three separate instances involved--one connected with the Winooski River near Montpelier, another attached to the West River in Windham County beyond Newfane, and a third centering in the Passumpsic in Caledonia County above Lyndonville. Of course many of the stray items mentioned other instances, but on analysis they all seemed to boil down to these three. In each case country folk reported seeing one or more very bizarre and disturbing objects in the surging waters that poured down from the unfrequented hills, and there was a widespread tendency to connect these sights with a primitive, half-forgotten cycle of whispered legend which old people resurrected for the occasion.

What people thought they saw were organic shapes not quite like any they had ever seen before. Naturally, there were many human bodies washed along by the streams in that tragic period; but those who described these strange shapes felt quite sure that they were not human, despite some superficial resemblances in size and general outline. Nor, said the witnesses, could they have been any kind of animal known to Vermont. They were pinkish things about five feet long; with crustaceous bodies bearing vast pairs of dorsal fins or membranous wings and several sets of articulated limbs, and with a sort of convoluted ellipsoid, covered with multitudes of very short antennae, where a head would ordinarily be. It was really remarkable how closely the reports from different sources tended to coincide; though the wonder was lessened by the fact that the old legends, shared at one time throughout the hill country, furnished a morbidly vivid picture which might well have coloured the imaginations of all the witnesses concerned. It was my conclusion that such witnesses--in every case naive and simple backwoods folk--had glimpsed the battered and bloated bodies of human beings or farm animals in the whirling currents; and had allowed the half-remembered folklore to invest these pitiful objects with fantastic attributes.

The ancient folklore, while cloudy, evasive, and largely forgotten by the present generation, was of a highly singular character, and obviously reflected the influence of still earlier Indian tales. I knew it well, though I had never been in Vermont, through the exceedingly rare monograph of Eli Davenport, which embraces material orally obtained prior to 1839 among the oldest people of the state. This material, moreover, closely coincided with tales which I had personally heard from elderly rustics in the mountains of New Hampshire. Briefly summarized, it hinted at a hidden race of monstrous beings which lurked somewhere among the remoter hills--in the deep woods of the highest peaks, and the dark valleys where streams trickle from unknown sources. These beings were seldom glimpsed, but evidences of their presence were reported by those who had ventured farther than usual up the slopes of certain mountains or into certain deep, steep-sided gorges that even the wolves shunned.

There were queer footprints or claw-prints in the mud of brook-margins and barren patches, and curious circles of stones, with the grass around them worn away, which did not seem to have been placed or entirely shaped by Nature. There were, too, certain caves of problematical depth in the sides of the hills; with mouths closed by boulders in a manner scarcely accidental, and with more than an average quota of the queer prints leading both toward and away from them--if indeed the direction of these prints could be justly estimated. And worst of all, there were the things which adventurous people had seen very rarely in the twilight of the remotest valleys and the dense perpendicular woods above the limits of normal hill-climbing.

It would have been less uncomfortable if the stray accounts of these things had not agreed so well. As it was, nearly all the rumors had several points in common; averring that the creatures were a sort of huge, light-red crab with many pairs of legs and with two great batlike wings in the middle of the back. They sometimes walked on all their legs, and sometimes on the hindmost pair only, using the others to convey large objects of indeterminate nature. On one occasion they were spied in considerable numbers, a detachment of them wading along a shallow woodland watercourse three abreast in evidently disciplined formation. Once a specimen was seen flying--launching itself from the top of a bald, lonely hill at night and vanishing in the sky after its great flapping wings had been silhouetted an instant against the full moon

These things seemed content, on the whole, to let mankind alone; though they were at times held responsible for the disappearance of venturesome individuals--especially persons who built houses too close to certain valleys or too high up on certain mountains. Many localities came to be known as inadvisable to settle in, the feeling persisting long after the cause was forgotten. People would look up at some of the neighbouring mountain-precipices with a shudder, even when not recalling how many settlers had been lost, and how many farmhouses burnt to ashes, on the lower slopes of those grim, green sentinels.

But while according to the earliest legends the creatures would appear to have harmed only those trespassing on their privacy; there were later accounts of their curiosity respecting men, and of their attempts to establish secret outposts in the human world. There were tales of the queer claw-prints seen around farmhouse windows in the morning, and of occasional disappearances in regions outside the obviously haunted areas. Tales, besides, of buzzing voices in imitation of human speech which made surprising offers to lone travelers on roads and cart-paths in the deep woods, and of children frightened out of their wits by things seen or heard where the primal forest pressed close upon their door-yards. In the final layer of legends--the layer just preceding the decline of superstition and the abandonment of close contact with the dreaded places--there are shocked references to hermits and remote farmers who at some period of life appeared to have undergone a repellent mental change, and who were shunned and whispered about as mortals who had sold themselves to the strange beings. In one of the northeastern counties it seemed to be a fashion about 1800 to accuse eccentric and unpopular recluses of being allies or representatives of the abhorred things.

As to what the things were--explanations naturally varied. The common name applied to them was "those ones," or "the old ones," though other terms had a local and transient use. Perhaps the bulk of the Puritan settlers set them down bluntly as familiars of the devil, and made them a basis of awed theological speculation. Those with Celtic legendry in their heritage--mainly the Scotch-Irish element of New Hampshire, and their kindred who had settled in Vermont on Governor Wentworth's colonial grants--linked them vaguely with the malign fairies and "little people" of the bogs and raths, and protected themselves with scraps of incantation handed down through many generations. But the Indians had the most fantastic theories of all. While different tribal legends differed, there was a marked consensus of belief in certain vital particulars; it being unanimously agreed that the creatures were not native to this earth.

The Pennacook myths, which were the most consistent and picturesque, taught that the Winged Ones came from the Great Bear in the sky, and had mines in our earthly hills whence they took a kind of stone they could not get on any other world. They did not live here, said the myths, but merely maintained outposts and flew back with vast cargoes of stone to their own stars in the north. They harmed only those earth-people who got too near them or spied upon them. Animals shunned them through instinctive hatred, not because of being hunted. They could not eat the things and animals of earth, but brought their own food from the stars. It was bad to get near them, and sometimes young hunters who went into their hills never came back. It was not good, either, to listen to what they whispered at night in the forest with voices like a bee's that tried to be like the voices of men. They knew the speech of all kinds of men--Pennacooks, Hurons, men of the Five Nations--but did not seem to have or need any speech of their own. They talked with their heads, which changed colour in different ways to mean different things.

All the legendry, of course, white and Indian alike, died down during the nineteenth century, except for occasional atavistical flareups. The ways of the Vermonters became settled; and once their habitual paths and dwellings were established according to a certain fixed plan, they remembered less and less what fears and avoidances had determined that plan, and even that there had been any fears or avoidances. Most people simply knew that certain hilly regions were considered as highly unhealthy, unprofitable, and generally unlucky to live in, and that the farther one kept from them the better off one usually was. In time the ruts of custom and economic interest became so deeply cut in approved places that there was no longer any reason for going outside them, and the haunted hills were left deserted by accident rather than by design. Save during infrequent local scares, only wonder-loving grandmothers and retrospective nonagenarians ever whispered of beings dwelling in those hills; and even such whispers admitted that there was not much to fear from those things now that they were used to the presence of houses and settlements, and now that human beings let their chosen territory severely alone.

All this I had long known from my reading, and from certain folk tales picked up in New Hampshire; hence when the flood-time rumours began to appear, I could easily guess what imaginative background had evolved them. I took great pains to explain this to my friends, and was correspondingly amused when several contentious souls continued to insist on a possible element of truth in the reports. Such persons tried to point out that the early legends had a significant persistence and uniformity, and that the virtually unexplored nature of the Vermont hills made it unwise to be dogmatic about what might or might not dwell among them; nor could they be silenced by my assurance that all the myths were of a well-known pattern common to most of mankind and determined by early phases of imaginative experience which always produced the same type of delusion.

It was of no use to demonstrate to such opponents that the Vermont myths differed but little in essence from those universal legends of natural personification which filled the ancient world with fauns and dryads and satyrs, suggested the kallikanzarai of modern Greece, and gave to wild Wales and Ireland their dark hints of strange, small, and terrible hidden races of troglodytes and burrowers. No use, either, to point out the even more startlingly similar belief of the Nepalese hill tribes in the dreaded Mi-Go or "Abominable Snow-Men" who lurk hideously amidst the ice and rock pinnacles of the Himalayan summits. When I brought up this evidence, my opponents turned it against me by claiming that it must imply some actual historicity for the ancient tales; that it must argue the real existence of some queer elder earth-race, driven to hiding after the advent and dominance of mankind, which might very conceivably have survived in reduced numbers to relatively recent times--or even to the present.

The more I laughed at such theories, the more these stubborn friends asseverated them; adding that even without the heritage of legend the recent reports were too clear, consistent, detailed, and sanely prosaic in manner of telling, to be completely ignored. Two or three fanatical extremists went so far as to hint at possible meanings in the ancient Indian tales which gave the hidden beings a nonterrestrial origin; citing the extravagant books of Charles Fort with their claims that voyagers from other worlds and outer space have often visited the earth. Most of my foes, however, were merely romanticists who insisted on trying to transfer to real life the fantastic lore of lurking "little people" made popular by the magnificent horror-fiction of Arthur Machen.

Chapter 2

As was only natural under the circumstances, this piquant debating finally got into print in the form of letters to the Arkham Advertiser; some of which were copied in the press of those Vermont regions whence the flood-stories came. The Rutland Herald gave half a page of extracts from the letters on both sides, while the Brattleboro Reformer reprinted one of my long historical and mythological summaries in full, with some accompanying comments in "The Pendrifter's" thoughtful column which supported and applauded my skeptical conclusions. By the spring of 1928 I was almost a well-known figure in Vermont, notwithstanding the fact that I had never set foot in the state. Then came the challenging letters from Henry Akeley which impressed me so profoundly, and which took me for the first and last time to that fascinating realm of crowded green precipices and muttering forest streams.

Most of what I know of Henry Wentworth Akeley was gathered by correspondence with his neighbours, and with his only son in California, after my experience in his lonely farmhouse. He was, I discovered, the last representative on his home soil of a long, locally distinguished line of jurists, administrators, and gentlemen-agriculturists. In him, however, the family mentally had veered away from practical affairs to pure scholarship; so that he had been a notable student of mathematics, astronomy, biology, anthropology, and folklore at the University of Vermont. I had never previously heard of him, and he did not give many autobiographical details in his communications; but from the first I saw he was a man of character, education, and intelligence, albeit a recluse with very little worldly sophistication.

Despite the incredible nature of what he claimed, I could not help at once taking Akeley more seriously than I had taken any of the other challengers of my views. For one thing, he was really close to the actual phenomena--visible and tangible--that he speculated so grotesquely about; and for another thing, he was amazingly willing to leave his conclusions in a tenative state like a true man of science. He had no personal preferences to advance, and was always guided by what he took to be solid evidence. Of course I began by considering him mistaken, but gave him credit for being intelligently mistaken; and at no time did I emulate some of his friends in attributing his ideas, and his fear of the lonely green hills, to insanity. I could see that there was a great deal to the man, and knew that what he reported must surely come from strange circumstance deserving investigation, however little it might have to do with the fantastic causes he assigned. Later on I received from him certain material proofs which placed the matter on a somewhat different and bewilderingly bizarre basis.

I cannot do better than transcribe in full, so far as is possible, the long letter in which Akeley introduced himself, and which formed such an important landmark in my own intellectual history. It is no longer in my possession, but my memory holds almost every word of its portentous message; and again I affirm my confidence in the sanity of the man who wrote it. Here is the text--a text which reached me in the cramped, archaic-looking scrawl of one who had obviously not mingled much with the world during his sedate, scholarly life. R.F.D. #2,

Townshend, Windham Co., Vermont.

May 5,1928

Albert N. Wilmarth, Esq.,

118 Saltonstall St.,

Arkham, Mass.

My Dear Sir: I have read with great interest the Brattleboro Reformer's reprint (Apr. 23, '28) of your letter on the recent stories of strange bodies seen floating in our flooded streams last fall, and on the curious folklore they so well agree with. It is easy to see why an outlander would take the position you take, and even why "Pendrifter" agrees with you. That is the attitude generally taken by educated persons both in and out of Vermont, and was my own attitude as a young man (I am now 57) before my studies, both general and in Davenport's book, led me to do some exploring in parts of the hills hereabouts not usually visited. I was directed toward such studies by the queer old tales I used to hear from elderly farmers of the more ignorant sort, but now I wish I had let the whole matter alone. I might say, with all proper modesty, that the subject of anthropology and folklore is by no means strange to me. I took a good deal of it at college, and am familiar with most of the standard authorities such as Tylor, Lubbock, Frazer, Quatrefages, Murray, Osborn, Keith, Boule, G. Elliott Smith, and so on. It is no news to me that tales of hidden races are as old as all mankind. I have seen the reprints of letters from you, and those agreeing with you, in the Rutland Herald, and guess I know about where your controversy stands at the present time. What I desire to say now is, that I am afraid your adversaries are nearer right than yourself, even though all reason seems to be on your side. They are nearer right than they realise themselves--for of course they go only by theory, and cannot know what I know. If I knew as little of the matter as they, I would feel justified in believing as they do. I would be wholly on your side. You can see that I am having a hard time getting to the point, probably because I really dread getting to the point; but the upshot of the matter is that I have certain evidence that monstrous things do indeed live in the woods on the high hills which nobody visits. I have not seen any of the things floating in the rivers, as reported, but I have seen things like them under circumstances I dread to repeat. I have seen footprints, and of late have seen them nearer my own home (I live in the old Akeley place south of Townshend Village, on the side of Dark Mountain) than I dare tell you now. And I have overheard voices in the woods at certain points that I will not even begin to describe on paper. At one place I heard them so much that I took a phonograph therewith a dictaphone attachment and wax blank--and I shall try to arrange to have you hear the record I got. I have run it on the machine for some of the old people up here, and one of the voices had nearly scared them paralysed by reason of its likeness to a certain voice (that buzzing voice in the woods which Davenport mentions) that their grandmothers have told about and mimicked for them. I know what most people think of a man who tells about "hearing voices"--but before you draw conclusions just listen to this record and ask some of the older backwoods people what they think of it. If you can account for it normally, very well; but there must be something behind it. Ex nihilo nihil fit, you know. Now my object in writing you is not to start an argument but to give you information which I think a man of your tastes will find deeply interesting. This is private. Publicly I am on your side, for certain things show me that it does not do for people to know too much about these matters. My own studies are now wholly private, and I would not think of saying anything to attract people's attention and cause them to visit the places I have explored. It is true--terribly true--that there are non-human creatures watching us all the time; with spies among us gathering information. It is from a wretched man who, if he was sane (as I think he was) was one of those spies, that I got a large part of my clues to the matter. He later killed himself, but I have reason to think there are others now. The things come from another planet, being able to live in interstellar space and fly through it on clumsy, powerful wings which have a way of resisting the aether but which are too poor at steering to be of much use in helping them about on earth. I will tell you about this later if you do not dismiss me at once as a madman. They come here to get metals from mines that go deep under the hills, and I think I know where they come from. They will not hurt us if we let them alone, but no one can say what will happen if we get too curious about them. Of course a good army of men could wipe out their mining colony. That is what they are afraid of. But if that happened, more would come from outside--any number of them. They could easily conquer the earth, but have not tried so far because they have not needed to. They would rather leave things as they are to save bother. I think they mean to get rid of me because of what I have discovered. There is a great black stone with unknown hieroglyphics half worn away which I found in the woods on Round Hill, east of here; and after I took it home everything became different. If they think I suspect too much they will either kill me or take me off the earth to where they come from. They like to take away men of learning once in a while, to keep informed on the state of things in the human world. This leads me to my secondary purpose in addressing you--namely, to urge you to hush up the present debate rather than give it more publicity. People must be kept away from these hills, and in order to effect this, their curiosity ought not to be aroused any further. Heaven knows there is peril enough anyway, with promoters and real estate men flooding Vermont with herds of summer people to overrun the wild places and cover the hills with cheap bungalows. I shall welcome further communication with you, and shall try to send you that phonograph record and black stone (which is so worn that photographs don't show much) by express if you are willing. I say "try" because I think those creatures have a way of tampering with things around here. There is a sullen furtive fellow named Brown, on a farm

Chapter 2

near the village, who I think is their spy. Little by little they are trying to cut me off from our world because I know too much about their world. They have the most amazing way of finding out what I do. You may not even get this letter. I think I shall have to leave this part of the country and go live with my son in San Diego, Cal., if things get any worse, but it is not easy to give up the place you were born in, and where your family has lived for six generations. Also, I would hardly dare sell this house to anybody now that the creatures have taken notice of it. They seem to be trying to get the black stone back and destroy the phonograph record, but I shall not let them if I can help it. My great police dogs always hold them back, for there are very few here as yet, and they are clumsy in getting about. As I have said, their wings are not much use for short flights on earth. I am on the very brink of deciphering that stone--in a very terrible way--and with your knowledge of folklore you may be able to supply the missing links enough to help me. I suppose you know all about the fearful myths antedating the coming of man to the earth--the Yog-Sothoth and Cthulhu cycles--which are hinted at in the Necronomicon. I had access to a copy of that once, and hear that you have one in your college library under lock and key. To conclude, Mr. Wilmarth, I think that with our respective studies we can be very useful to each other. I don't wish to put you in any peril, and suppose I ought to warn you that possession of the stone and the record won't be very safe; but I think you will find any risks worth running for the sake of knowledge. I will drive down to Newfane or Brattleboro to send whatever you authorize me to send, for the express offices there are more to be trusted. I might say that I live quite alone now, since I can't keep hired help any more. They won't stay because of the things that try to get near the house at night, and that keep the dogs barking continually. I am glad I didn't get as deep as this into the business while my wife was alive, for it would have driven her mad. Hoping that I am not bothering you unduly, and that you will decide to get in touch with me rather than throw this letter into the waste basket as a madman's raving, I am Yrs. very truly, Henry W. Akeley P.S. I am making some extra prints of certain photographs taken by me, which I think will help to prove a number of the points I have touched on. The old people think they are monstrously true. I shall send you these very soon if you are interested.

H. W. A.

Chapter 2

It would be difficult to describe my sentiments upon reading this strange document for the first time. By all ordinary rules, I ought to have laughed more loudly at these extravagances than at the far milder theories which had previously moved me to mirth; yet something in the tone of the letter made me take it with paradoxical seriousness. Not that I believed for a moment in the hidden race from the stars which my correspondent spoke of; but that, after some grave preliminary doubts, I grew to feel oddly sure of his sanity and sincerity, and of his confrontation by some genuine though singular and abnormal phenomenon which he could not explain except in this imaginative way. It could not be as he thought it, I reflected, yet on the other hand, it could not be otherwise than worthy of investigation. The man seemed unduly excited and alarmed about something, but it was hard to think that all cause was lacking. He was so specific and logical in certain ways--and after all, his yarn did fit in so perplexingly well with some of the old myths--even the wildest Indian legends.

That he had really overheard disturbing voices in the hills, and had really found the black stone he spoke about, was wholly possible despite the crazy inferences he had made--inferences probably suggested by the man who had claimed to be a spy of the outer beings and had later killed himself. It was easy to deduce that this man must have been wholly insane, but that he probably had a streak of perverse outward logic which made the naive Akeley--already prepared for such things by his folklore studies--believe his tale. As for the latest developments--it appeared from his inability to keep hired help that Akeley's humbler rustic neighbours were as convinced as he that his house was besieged by uncanny things at night. The dogs really barked, too.

And then the matter of that phonograph record, which I could not but believe he had obtained in the way he said. It must mean something; whether animal noises deceptively like human speech, or the speech of some hidden, night-haunting human being decayed to a state not much above that of lower animals. From this my thoughts went back to the black hieroglyphed stone, and to speculations upon what it might mean. Then, too, what of the photographs which Akeley said he was about to send, and which the old people had found so convincingly terrible?

Chapter 2

As I re-read the cramped handwriting I felt as never before that my credulous opponents might have more on their side than I had conceded. After all, there might be some queer and perhaps hereditarily misshapen outcasts in those shunned hills, even though no such race of star-born monsters as folklore claimed. And if there were, then the presence of strange bodies in the flooded streams would not be wholly beyond belief. Was it too presumptuous to suppose that both the old legends and the recent reports had this much of reality behind them? But even as I harboured these doubts I felt ashamed that so fantastic a piece of bizarrerie as Henry Akeley's wild letter had brought them up.

In the end I answered Akeley's letter, adopting a tone of friendly interest and soliciting further particulars. His reply came almost by return mail; and contained, true to promise, a number of Kodak views of scenes and objects illustrating what he had to tell. Glancing at these pictures as I took them from the envelope, I felt a curious sense of fright and nearness to forbidden things; for in spite of the vagueness of most of them, they had a damnably suggestive power which was intensified by the fact of their being genuine photographs--actual optical links with what they portrayed, and the product of an impersonal transmitting process without prejudice, fallibility, or mendacity.

The more I looked at them, the more I saw that my senous estimate of Akeley and his story had not been unjustified. Certainly, these pictures carried conclusive evidence of something in the Vermont hills which was at least vastly outside the radius of our common knowledge and belief. The worst thing of all was the footprint--a view taken where the sun shone on a mud patch somewhere in a deserted upland. This was no cheaply counterfeited thing, I could see at a glance; for the sharply defined pebbles and grassblades in the field of vision gave a clear index of scale and left no possibility of a tricky double exposure. I have called the thing a "footprint," but "claw-print" would be a better term. Even now I can scarcely describe it save to say that it was hideously crablike, and that there seemed to be some ambiguity about its direction. It was not a very deep or fresh print, but seemed to be about the size of an average man's foot. From a central pad, pairs of saw-toothed nippers projected in opposite directions--quite baffling as to function, if indeed the whole object were exclusively an organ of locomotion.

Another photograph--evidently a time-exposure taken in deep shadow--was of the mouth of a woodland cave, with a boulder of, rounded regularity choking the aperture. On the bare ground in front of it, one could just discern a dense network of curious tracks, and when I studied the picture with a magnifier I felt uneasily sure that the tracks were like the one in the other view. A third pictured showed a druid-like circle of standing stones on the summit of a wild hill. Around the cryptic circle the grass was very much beaten down and worn away, though I could not detect any footprints even with the glass. The extreme remoteness of the place was apparent from the veritable sea of tenantless mountains which formed the background and stretched away toward a misty horizon.

But if the most disturbing of all the views was that of the footprint, the most curiously suggestive was that of the great black stone found in the Round Hill woods. Akeley had photographed it on what was evidently his study table, for I could see rows of books and a bust of Milton in the background. The thing, as nearly as one might guess, had faced the camera vertically with a somewhat irregularly curved surface of one by two feet; but to say anything definite about that surface, or about the general shape of the whole mass, almost defies the power of language. What outlandish geometrical principles had guided its cutting--for artificially cut it surely was--I could not even begin to guess; and never before had I seen anything which struck me as so strangely and unmistakably alien to this world. Of the hieroglyphics on the surface I could discern very few, but one or two that I did see gave rather a shock. Of course they might be fraudulent, for others besides myself had read the monstrous and abhorred Necronomicon of the mad Arab Abdul Alhazred; but it nevertheless made me shiver to recognise certain ideographs which study had taught me to link with the most blood-curdling and blasphemous whispers of things that had had a kind of mad half-existence before the earth and the other inner worlds of the solar system were made.

Of the five remaining pictures, three were of swamp and hill scenes which seemed to bear traces of hidden and unwholesome tenancy.

Another was of a queer mark in the ground very near Akeley's house, which he said he had photographed the morning after a night on which the dogs had barked more violently than usual. It was very blurred, and one could really draw no certain conclusions from it; but it did seem fiendishly like that other mark or claw-print photographed on the deserted upland. The final picture was of the Akeley place itself: a trim white house of two stories and attic, about a century and a quarter old, and with a well-kept lawn and stone-bordered path leading up to a tastefully carved Georgian doorway. There were several huge police dogs on the lawn, squatting near a pleasant-faced man with a close-cropped grey beard whom I took to be Akeley himself--his own photographer, one might infer from the tube-connected bulb in his right hand.

From the pictures I turned to the bulky, closely-written letter itself; and for the next three hours was immersed in a gulf of unutterable horror. Where Akeley had given only outlines before, he now entered into minute details; presenting long transcripts of words overheard in the woods at night, long accounts of monstrous pinkish forms spied in thickets at twilight on the hills, and a terrible cosmic narrative derived from the application of profound and varied scholarship to the endless bygone discourses of the mad self-styled spy who had killed himself. I found myself faced by names and terms that I had heard elsewhere in the most hideous of connections--Yuggoth, Great Cthulhu, Tsathoggua, YogSothoth, R'lyeh, Nyarlathotep, Azathoth, Hastur, Yian, Leng, the Lake of Hali, Bethmoora, the Yellow Sign, L'mur-Kathulos, Bran, and the Magnum Innominandum--and was drawn back through nameless aeons and inconceivable dimensions to worlds of elder, outer entity at which the crazed author of the Necronomicon had only guessed in the vaguest way. I was told of the pits of primal life, and of the streams that had trickled down therefrom; and finally, of the tiny rivulets from one of those streams which had become entangled with the destinies of our own earth.

My brain whirled; and where before I had attempted to explain things away, I now began to believe in the most abnormal and incredible wonders. The array of vital evidence was damnably vast and overwhelming; and the cool, scientific attitude of Akeley--an attitude removed as far as imaginable from the demented, the fanatical, the hysterical, or even the. extravagantly speculative--had a tremendous effect on my thought and judgment. By the time I laid the frightful letter aside I could understand the fears he had come to entertain, and was ready to do anything in my power to keep people away from those wild, haunted hills. Even now, when time has dulled the impression and made me half-question my own experience and horrible doubts, there are things in that letter of Akeley's which I would not quote, or even form into words on paper. I am almost glad that the letter and record and photographs are gone now--and I wish, for reasons I shall soon make clear, that the new planet beyond Neptune had not been discovered.

With the reading of that letter my public debating about the Vermont horror permanently ended. Arguments from opponents remained unanswered or put off with promises, and eventually the controversy petered out into oblivion. During late May and June I was in constant correspondence with Akeley; though once in a while a letter would be lost, so that we would have to retrace our ground and perform considerable laborious copying. What we were trying to do, as a whole, was to compare notes in matters of obscure mythological scholarship and arrive at a clearer correlation of the Vermont horrors with the general body of primitive world legend.

For one thing, we virtually decided that these morbidities and the hellish Himalayan Mi-Go were one and the same order of incarnated nightmare. There was also absorbing zoological conjectures, which I would have referred to Professor Dexter in my own college but for Akeley's imperative command to tell no one of the matter before us. If I seem to disobey that command now, it is only because I think that at this stage a warning about those farther Vermont hills--and about those Himalayan peaks which bold explorers are more and more determined to ascend--is more conducive to public safety than silence would be. One specific thing we were leading up to was a deciphering of the hieroglyphics on that infamous black stone--a deciphering which might well place us in possession of secrets deeper and more dizzying than any formerly known to man.

Chapter 3

Toward the end of June the phonograph record came--shipped from Brattleboro, since Akeley was unwilling to trust conditions on the branch line north of there. He had begun to feel an increased sense of espionage, aggravated by the loss of some of our letters; and said much about the insidious deeds of certain men whom he considered tools and agents of the hidden beings. Most of all he suspected the surly farmer Walter Brown, who lived alone on a run-down hillside place near the deep woods, and who was often seen loafing around corners in Brattleboro, Bellows Falls, Newfane, and South Londonderry in the most inexplicable and seemingly unmotivated way. Brown's voice, he felt convinced, was one of those he had overheard on a certain occasion in a very terrible conversation; and he had once found a footprint or clawprint near Brown's house which might possess the most ominous significance. It had been curiously near some of Brown's own footprints--footprints that faced toward it.

So the record was shipped from Brattleboro, whither Akeley drove in his Ford car along the lonely Vermont back roads. He confessed in an accompanying note that he was beginning to be afraid of those roads, and that he would not even go into Townshend for supplies now except in broad daylight. It did not pay, he repeated again and again, to know too much unless one were very remote from those silent and problematical hills. He would be going to California pretty soon to live with his son, though it was hard to leave a place where all one's memories and ancestral feelings centered.

Before trying the record on the commercial machine which I borrowed from the college administration building I carefully went over all the explanatory matter in Akeley's various letters. This record, he had said, was obtained about 1 A.M. on the 1st of May, 1915, near the closed mouth of a cave where the wooded west slope of Dark Mountain rises out of Lee's swamp. The place had always been unusually plagued with strange voices, this being the reason he had brought the phonograph, dictaphone, and blank in expectation of results. Former experience had told him that May Eve--the hideous Sabbat-night of underground European legend--would probably be more fruitful than any other date, and he was not disappointed. It was noteworthy, though, that he never again heard voices at that particular spot.

Unlike most of the overheard forest voices, the substance of the record was quasi-ritualistic, and included one palpably human voice which Akeley had never been able to place. It was not Brown's, but seemed to be that of a man of greater cultivation. The second voice, however, was the real crux of the thing--for this was the accursed buzzing which had no likeness to humanity despite the human words which it uttered in good English grammar and a scholarly accent.

The recording phonograph and dictaphone had not worked uniformly well, and had of course been at a great disadvantage because of the remote and muffled nature of the overheard ritual; so that the actual speech secured was very fragmentary. Akeley had given me a transcript of what he believed the spoken words to be, and I glanced through this again as I prepared the machine for action. The text was darkly mysterious rather than openly horrible, though a knowledge of its origin and manner of gathering gave it all the associative horror which any words could well possess. I will present it here in full as I remember it--and I am fairly confident that I know it correctly by heart, not only from reading the transcript, but from playing the record itself over and over again. It is not a thing which one might readily forget! (Indistinguishable Sounds)

(A Cultivated Male Human Voice)

...is the Lord of the Wood, even to... and the gifts of the men of Leng... so from the wells of night to the gulfs of space, and from the gulfs of space to the wells of night, ever the praises of Great Cthulhu, of Tsathoggua, and of Him Who is not to be Named. Ever Their praises, and abundance to the Black Goat of the Woods. Ia! Shub-Niggurath! The Goat with a Thousand Young!

(A Buzzing Imitation of Human Speech)

Ia! Shub-Niggurath! The Black Goat of the Woods with a Thousand Young!

(Human Voice)

Chapter 3

And it has come to pass that the Lord of the Woods, being... seven and nine, down the onyx steps... (tri)butes to Him in the Gulf, Azathoth, He of Whom Thou has taught us marv(els)... on the wings of night out beyond space, out beyond th... to That whereof Yuggoth is the youngest child, rolling alone in black aether at the rim...

(Buzzing Voice)

...go out among men and find the ways thereof, that He in the Gulf may know. To Nyarlathotep, Mighty Messenger, must all things be told. And He shall put on the semblance of men, the waxen mask and the robe that hides, and come down from the world of Seven Suns to mock...

(Human Voice)

(Nyarl)athotep, Great Messenger, bringer of strange joy to Yuggoth through the void, Father of the Million Favoured Ones, Stalker among...

(Speech Cut Off by End of Record)

Such were the words for which I was to listen when I started the phonograph. It was with a trace of genuine dread and reluctance that I pressed the lever and heard the preliminary scratching of the sapphire point, and I was glad that the first faint, fragmentary words were in a human voice--a mellow, educated voice which seemed vaguely Bostonian in accent, and which was certainly not that of any native of the Vermont hills. As I listened to the tantalisingly feeble rendering, I seemed to find the speech identical with Akeley's carefully prepared transcript. On it chanted, in that mellow Bostonian voice... "Ia! Shub-Niggurath! The Goat with a Thousand Young!..."

And then I heard the other voice. To this hour I shudder retrospectively when I think of how it struck me, prepared though I was by Akeley's accounts. Those to whom I have since described the record profess to find nothing but cheap imposture or madness in it; but could they have the accursed thing itself, or read the bulk of Akeley's correspondence, (especially that terrible and encyclopaedic second letter), I know they would think differently. It is, after all, a tremendous pity that I did not disobey Akeley and play the record for others--a tremendous pity, too, that all of his letters were lost. To me, with my first-hand impression of the actual sounds, and with my knowledge of the background and surrounding circumstances, the voice was a monstrous thing. It swiftly followed the human voice in ritualistic response, but in my imagination it was a morbid echo winging its way across unimaginable abysses from unimaginable outer hells. It is more than two years now since I last ran off that blasphemous waxen cylinder; but at this moment, and at all other moments, I can still hear that feeble, fiendish buzzing as it reached me for the first time.

"Ia! Shub-Niggurath! The Black Goat of the Woods with a Thousand Young!"

But though the voice is always in my ears, I have not even yet been able to analyse it well enough for a graphic description. It was like the drone of some loathsome, gigantic insect ponderously shaped into the articulate speech of an alien species, and I am perfectly certain that the organs producing it can have no resemblance to the vocal organs of man, or indeed to those of any of the mammalia. There were singularities in timbre, range, and overtones which placed this phenomenon wholly outside the sphere of humanity and earth-life. Its sudden advent that first time almost stunned me, and I heard the rest of the record through in a sort of abstracted daze. When the longer passage of buzzing came, there was a sharp intensification of that feeling of blasphemous infinity which had struck me during the shorter and earlier passage. At last the record ended abruptly, during an unusually clear speech of the human and Bostonian voice; but I sat stupidly staring long after the machine had automatically stopped.

I hardly need say that I gave that shocking record many another playing, and that I made exhaustive attempts at analysis and comment in comparing notes with Akeley. It would be both useless and disturbing to repeat here all that we concluded; but I may hint that we agreed in believing we had secured a clue to the source of some of the most repulsive primordial customs in the cryptic elder religions of mankind. It seemed plain to us, also, that there were ancient and elaborate alliance; between the hidden outer creatures and certain members of the human race. How extensive these alliances were, and how their state today might compare with their state in earlier ages, we had no means of guessing; yet at best there was room for a limitless amount of horrified speculation. There seemed to be an awful, immemorial linkage in several definite stages betwixt man and nameless infinity. The blasphemies which appeared on earth, it was hinted, came from the dark planet Yuggoth, at the rim of the solar system; but this was itself merely the populous outpost of a frightful interstellar race whose ultimate source must lie far outside even the Einsteinian space-time continuum or greatest known cosmos.

Meanwhile we continued to discuss the black stone and the best way of getting it to Arkham--Akeley deeming it inadvisable to have me visit him at the scene of his nightmare studies. For some reason or other, Akeley was afraid to trust the thing to any ordinary or expected transportation route. His final idea was to take it across country to Bellows Falls and ship it on the Boston and Maine system through Keene and Winchendon and Fitchburg, even though this would necessitate his driving along somewhat lonelier and more forest-traversing hill roads than the main highway to Brattleboro. He said he had noticed a man around the express office at Brattleboro when he had sent the phonograph record, whose actions and expression had been far from reassuring. This man had seemed too anxious to talk with the clerks, and had taken the train on which the record was shipped. Akeley confessed that he had not felt strictly at ease about that record until he heard from me of its safe receipt.

About this time--the second week in July--another letter of mine went astray, as I learned through an anxious communication from Akeley. After that he told me to address him no more at Townshend, but to send all mail in care of the General Delivery at Brattleboro; whither he would make frequent trips either in his car or on the motor-coach line which had lately replaced passenger service on the lagging branch railway. I could see that he was getting more and more anxious, for he went into much detail about the increased barking of the dogs on moonless nights, and about the fresh claw-prints he sometimes found in the road and in the mud at the back of his farmyard when morning came. Once he told about a veritable army of prints drawn up in a line facing an equally thick and resolute line of dog-tracks, and sent a loathsomely disturbing Kodak picture to prove it. That was after a night on which the dogs had outdone themselves in barking and howling.

On the morning of Wednesday, July 18, I received a telegram from Bellows Falls, in which Akeley said he was expressing the black stone over the B. & M. on Train No. 5508, leaving Bellows Falls at 12:15 P.M., standard time, and due at the North Station in Boston at 4:12 P.M. It ought, I calculated, to get up to Arkham at least by the next noon; and accordingly I stayed in all Thursday morning to receive it. But noon came and went without its advent, and when I telephoned down to the express office I was informed that no shipment for me had arrived. My next act, performed amidst a growing alarm, was to give a long-distance call to the express agent at the Boston North Station; and I was scarcely surprised to learn that my consignment had not appeared. Train No. 5508 had pulled in only 35 minutes late on the day before, but had contained no box addressed to me. The agent promised, however, to institute a searching inquiry; and I ended the day by sending Akeley a night-letter outlining the situation.

With commendable promptness a report came from the Boston office on the following afternoon, the agent telephoning as soon as he learned the facts. It seemed that the railway express clerk on No. 5508 had been able to recall an incident which might have much bearing on my loss--an argument with a very curious-voiced man, lean, sandy, and rustic-looking, when the train was waiting at Keene, N. H., shortly after one o'clock standard time. The man, he said, was greatly excited about a heavy box which he claimed to expect, but which was neither on the train nor entered on the company's books. He had given the name of Stanley Adams, and had had such a queerly thick droning voice, that it made the clerk abnormally dizzy and sleepy to listen to him. The clerk could not remember quite how the conversation had ended, but recalled starting into a fuller awakeness when the train began to move. The Boston agent added that this clerk was a young man of wholly unquestioned veracity and reliability, of known antecedents and long with the company.

That evening I went to Boston to interview the clerk in person, having obtained his name and address from the office. He was a frank, prepossessing fellow, but I saw that he could add nothing to his original account. Oddly, he was scarcely sure that he could even recognise the strange inquirer again. Realising that he had no more to tell, I returned to Arkham and sat up till morning writing letters to Akeley, to the express company and to the police department and station agent in Keene. I felt that the strange-voiced man who had so queerly affected the clerk must have a pivotal place in the ominous business, and hoped that Keene station employees and telegraph-office records might tell something about him and about how he happened to make his inquiry when and where he did.

I must admit, however, that all my investigations came to nothing. The queer-voiced man had indeed been noticed around the Keene station in the early afternoon of July 18, and one lounger seemed to couple him vaguely with a heavy box; but he was altogether unknown, and had not been seen before or since. He had not visited the telegraph office or received any message so far as could be learned, nor had any message which might justly be considered a notice of the black stone's presence on No. 5508 come through the office for anyone. Naturally Akeley joined with me in conducting these inquiries, and even made a personal trip to Keene to question the people around the station; but his attitude toward the matter was more fatalistic than mine. He seemed to find the loss of the box a portentous and menacing fulfillment of inevitable tendencies, and had no real hope at all of its recovery. He spoke of the undoubted telepathic and hypnotic powers of the hill creatures and their agents, and in one letter hinted that he did not believe the stone was on this earth any longer. For my part, I was duly enraged, for I had felt there was at least a chance of learning profound and astonishing things from the old, blurred hieroglyphs. The matter would have rankled bitterly in my mind had not Akeley's immediately subsequent letters brought up a new phase of the whole horrible hill problem which at once seized all my attention.

Chapter 4

The unknown things, Akeley wrote in a script grown pitifully tremulous, had begun to close in on him with a wholly new degree of determination. The nocturnal barking of the dogs whenever the moon. was dim or absent was hideous now, and there had been attempts to molest him on the lonely roads he had to traverse by day. On the second of August, while bound for the village in his car, he had found a tree-trunk laid in his path at a point where the highway ran through a deep patch of woods; while the savage barking of the two great dogs he had with him told all too well of the things which must have been lurking near. What would have happened had the dogs not been there, he did not dare guess--but he never went out now without at least two of his faithful and powerful pack. Other road experiences had occurred on August fifth and sixth; a shot grazing his car on one occasion, and the barking of the dogs telling of unholy woodland presences on the other.

On August fifteenth I received a frantic letter which disturbed me greatly, and which made me wish Akeley could put aside his lonely reticence and call in the aid of the law. There had been frightful happening on the night of the 12-13th, bullets flying outside the farmhouse, and three of the twelve great dogs being found shot dead in the morning. There were myriads of claw-prints in the road, with the human prints of Walter Brown among them. Akeley had started to telephone to Brattleboro for more dogs, but the wire had gone dead before he had a chance to say much. Later he went to Brattleboro in his car, and learned there that linemen had found the main cable neatly cut at a point where it ran through the deserted hills north of Newfane. But he was about to start home with four fine new dogs, and several cases of ammunition for his big-game repeating rifle. The letter was written at the post office in Brattleboro, and came through to me without delay.

My attitude toward the matter was by this time quickly slipping from a scientific to an alarmedly personal one. I was afraid for Akeley in his remote, lonely farmhouse, and half afraid for myself because of my now definite connection with the strange hill problem. The thing was reaching out so. Would it suck me in and engulf me? In replying to his letter I urged him to seek help, and hinted that I might take action myself if he did not. I spoke of visiting Vermont in person in spite of his wishes, and of helping him explain the situation to the proper authorities. In return, however, I received only a telegram from Bellows Falls which read thus:

APPRECIATE YOUR POSITION BUT CAN DO NOTHING TAKE NO ACTION YOURSELF FOR IT COULD ONLY HARM BOTH WAIT FOR EXPLANATION HENRY AKELY

But the affair was steadily deepening. Upon my replying to the telegram I received a shaky note from Akeley with the astonishing news that he had not only never sent the wire, but had not received the letter from me to which it was an obvious reply. Hasty inquiries by him at Bellows Falls had brought out that the message was deposited by a strange sandy-haired man with a curiously thick, droning voice, though more than this he could not learn. The clerk showed him the original text as scrawled in pencil by the sender, but the handwriting was wholly unfamiliar. It was noticeable that the signature was misspelled--A-K-E-L-Y, without the second "E." Certain conjectures were inevitable, but amidst the obvious crisis he did not stop to elaborate upon them,

He spoke of the death of more dogs and the purchase of still others, and of the exchange of gunfire which had become a settled feature each moonless night. Brown's prints, and the prints of at least one or two more shod human figures, were now found regularly among the claw-prints in the road, and at the back of the farmyard. It was, Akeley admitted, a pretty bad business; and before long he would probably have to go to live with his California son whether or not he could sell the old place. But it was not easy to leave the only spot one could really think of as home. He must try to hang on a little longer; perhaps he could scare off the intruders--especially if he openly gave up all further attempts to penetrate their secrets.

Writing Akeley at once, I renewed my offers of aid, and spoke again of visiting him and helping him convince the authorities of his dire peril. In his reply he seemed less set against that plan than his past attitude would have led one to predict, but said he would like to hold off a little while longer--long enough to get his things in order and reconcile himself to the idea of leaving an almost morbidly cherished birthplace. People looked askance at his studies and speculations and it would be better to get quietly off without setting the countryside in a turmoil and creating widespread doubts of his own sanity. He had had enough, he admitted, but he wanted to make a dignified exit if he could.

This letter reached me on the 28th of August, and I prepared and mailed as encouraging a reply as I could. Apparently the encouragement had effect, for Akeley had fewer terrors to report when he acknowledged my note. He was not very optimistic, though, and expressed the belief that it was only the full moon season which was holding the creatures off. He hoped there would not be many densely cloudy nights, and talked vaguely of boarding in Brattleboro when the moon waned. Again I wrote him encouragingly but on September 5th there came a fresh communication which had obviously crossed my letter in the mails; and to this I could not give any such hopeful response. In view of its importance I believe I had better give it in full--as best I can do from memory of the shaky script. It ran substantially as follows:

Monday

Dear Wilmarth

A rather discouraging P. S. to my last. Last night was thickly cloudy--though no rain--and not a bit of moonlight got through. Things were pretty bad, and I think the end is getting near, in spite of all we have hoped. After midnight something landed on the roof of the house, and the dogs all rushed up to see what it was. I could hear them snapping and tearing around, and then one managed to get on the roof by jumping from the low ell. There was a terrible fight up there, and I heard a frightful buzzing which I'll never forget. And then there was a shocking smell. About the same time bullets came through the window and nearly grazed me. I think the main line of the hill creatures had got close to the house when the dogs divided because of the roof business. What was up there I don't know yet, but I'm afraid the creatures are learning to steer better with their space wings. I put out the light and used the windows for loopholes, and raked all around the house with rifle fire aimed just high enough not to hit the dogs. That seemed to end the business, but in the morning I found great pools of blood in the yard, besides pools of a green sticky stuff that had the worst odour I have ever smelled. I climbed up on the roof and found more of the sticky stuff there. Five of the dogs were killed--I'm afraid I hit one myself by aiming too low, for he was shot in the back. Now I am setting the panes the shots broke, and am going to Brattleboro for more dogs. I guess the men at the kennels think I am crazy. Will drop another note later. Suppose I'll be ready for moving in a week or two, though it nearly kills me to think of it.

Hastily--Akeley

But this was not the only letter from Akeley to cross mine. On the next morning--September 6th--still another came; this time a frantic scrawl which utterly unnerved me and put me at a loss what to say or do next. Again I cannot do better than quote the text as faithfully as memory will let me.

Tuesday Clouds didn't break, so no moon again--and going into the wane anyhow. I'd have the house wired for electricity and put in a searchlight if I didn't know they'd cut the cables as fast as they could be mended. I think I am going crazy. It may be that all I have ever written you is a dream or madness. It was bad enough before, but this time it is too much. They talked to me last night--talked in that cursed buzzing voice and told me things that I dare not repeat to you. I heard them plainly above the barking of the dogs, and once when they were drowned out a human voice helped them. Keep out of this, Wilmarth--it is worse than either you or I ever suspected. They don't mean to let me get to California now--they want to take me off alive, or what theoretically and mentally amounts to alive--not only to Yuggoth, but beyond that--away outside the galaxy and possibly beyond the last curved rim of space. I told them I wouldn't go where they wish, or in the terrible way they propose to take me, but I'm afraid it will be no use. My place is so far out that they may come by day as well as by night before long. Six more dogs killed, and I felt presences all along the wooded parts of the road when I drove to Brattleboro today. It was a mistake for me to try to send you that phonograph record and black stone. Better smash the record before it's too late. Will drop you another line tomorrow if I'm still here. Wish I could arrange to get my books and things to Brattleboro and board there. I would run off without anything if I could but something inside my mind holds me back. I can slip out to Brattleboro, where I ought to be safe, but I feel just as much a prisoner there as at the house. And I seem to know that I couldn't get much farther even if I dropped everything and tried. It is horrible--don't get mixed up in this. Yrs--Akeley

I did not sleep at all the night after receiving this terrible thing, and was utterly baffled as to Akeley's remaining degree of sanity. The substance of the note was wholly insane, yet the manner of expression--in view of all that had gone before--had a grimly potent quality of convincingness. I made no attempt to answer it, thinking it better to wait until Akeley might have time to reply to my latest communication. Such a reply indeed came on the following day, though the fresh material in it quite overshadowed any of the points brought up by the letter nominally answered. Here is what I recall of the text, scrawled and blotted as it was in the course of a plainly frantic and hurried composition.

Wednesday

W--Your letter came, but it's no use to discuss anything any more. I am fully resigned. Wonder that I have even enough will power left to fight them off. Can't escape even if I were willing to give up everything and run. They'll get me. Had a letter from them yesterday--R.F.D. man brought it while I was at Brattleboro. Typed and postmarked Bellows Falls. Tells what they want to do with me--I can't repeat it. Look out for yourself, too! Smash that record. Cloudy nights keep up, and moon waning all the time. Wish I dared to get help--it might brace up my will power--but everyone who would dare to come at all would call me crazy unless there happened to be some proof. Couldn't ask people to come for no reason at all--am all out of touch with everybody and have been for years. But I haven't told you the worst, Wilmarth. Brace up to read this, for it will give you a shock. I am telling the truth, though. It is this--I have seen and touched one of the things, or part of one of the things. God, man, but it's awful! It was dead, of course. One of the dogs had it, and I found it near the kennel this morning. I tried to save it in the woodshed to convince people of the whole thing, but it all evaporated in a few hours. Nothing left. You know, all those things in the rivers were seen only on the first morning after the flood. And here's the worst. I tried to photograph it for you, but when I developed the film there wasn't anything visible except the woodshed. What can the thing have been made of? I saw it and felt it, and they all leave footprints. It was surely made of matter--but what kind of matter? The shape can't be described. It was a great crab with a lot of pyramided fleshy rings or knots of thick, ropy stuff covered with feelers where a man's head would be. That green sticky stuff is its blood or juice. And there are more of them due on earth any minute. Walter Brown is missing--hasn't been seen loafing around any of his usual corners in the villages hereabouts. I must have got him with one of my shots, though the creatures always seem to try to take their dead and wounded away. Got into town this afternoon without any trouble, but am afraid they're beginning to hold off because they're sure of me. Am writing this in Brattleboro P. 0. This may be goodbye--if it is, write my son George Goodenough Akeley, 176 Pleasant St., San Diego, Cal., but don't come up here. Write the boy if you don't hear from me in a week, and watch the papers for news. I'm going to play my last two cards now--if I have the will power left. First to try poison gas on the things (I've got the right chemicals and have fixed up masks for myself and the dogs) and then if that doesn't work, tell the sheriff. They can lock me in a madhouse if they want to--it'll be better than what the other creatures would do. Perhaps I can get them to pay attention to the prints around the house--they are faint, but I can find them every morning. Suppose, though, police would say I faked them somehow; for they all think I'm a queer character. Must try to have a state policeman spend a night here and see for himself--though it would be just like the creatures to learn about it and hold off that night. They cut my wires whenever I try to telephone in the night--the linemen think it is very queer, and may testify for me if they don't go and imagine I cut them myself. I haven't tried to keep them repaired for over a week now. I could get some of the ignorant people to testify for me about the reality of the horrors, but everybody laughs at what they say, and anyway, they have shunned my place for so long that they don't know any of the new events. You couldn't get one of those rundown farmers to come within a mile of my house for love or money. The mail-carrier hears what they say and jokes me about it--God! If I only dared tell him how real it is! I think I'll try to get him to notice the prints, but he comes

Chapter 4

in the afternoon and they're usually about gone by that time. If I kept one by setting a box or pan over it, he'd think surely it was a fake or joke. Wish I hadn't gotten to be such a hermit, so folks don't drop around as they used to. I've never dared show the black stone or the Kodak pictures, or play that record, to anybody but the ignorant people. The others would say I faked the whole business and do nothing but laugh. But I may yet try showing the pictures. They give those claw-prints clearly, even if the things that made them can't be photographed. What a shame nobody else saw that thing this morning before it went to nothing! But I don't know as I care. After what I've been through, a madhouse is as good a place as any. The doctors can help me make up my mind to get away from this house, and that is all that will save me. Write my son George if you don't hear soon. Goodbye, smash that record, and don't mix up in this.

Yrs--Akeley

This letter frankly plunged me into the blackest of terror. I did not know what to say in answer, but scratched off some incoherent words of advice and encouragement and sent them by registered mail. I recall urging Akeley to move to Brattleboro at once, and place himself under the protection of the authorities; adding that I would come to that town with the phonograph record and help convince the courts of his sanity. It was time, too, I think I wrote, to alarm the people generally against this thing in their midst. It will be observed that at this moment of stress my own belief in all Akeley had told and claimed was virtually complete, though I did think his failure to get a picture of the dead monster was due not to any freak of Nature but to some excited slip of his own.

Chapter 5

Then, apparently crossing my incoherent note and reaching me Saturday afternoon, September 8th, came that curiously different and calming letter neatly typed on a new machine; that strange letter of reassurance and invitation which must have marked so prodigious a transition in the whole nightmare drama of the lonely hills. Again I will quote from memory--seeking for special reasons to preserve as much of the flavour of the style as I can. It was postmarked Bellows Falls, and the signature as well as the body of the letter was typed--as is frequent with beginners in typing. The text, though, was marvellously accurate for a tyro's work; and I concluded that Akeley must have used a machine at some previous period--perhaps in college. To say that the letter relieved me would be only fair, yet beneath my relief lay a substratum of uneasiness. If Akeley had been sane in his terror, was he now sane in his deliverance? And the sort of "improved rapport" mentioned... what was it? The entire thing implied such a diametrical reversal of Akeley's previous attitude! But here is the substance of the text, carefully transcribed from a memory in which I take some pride.

Townshend, Vermont,

Thursday, Sept. 6, 1928.

My dear Wilmarth:--It gives me great pleasure to be able to set you at rest regarding all the silly things I've been writing you. I say "silly," although by that I mean my frightened attitude rather than my descriptions of certain phenomena. Those phenomena are real and important enough; my mistake had been in establishing an anomalous attitude toward them. I think I mentioned that my strange visitors were beginning to communicate with me, and to attempt such communication. Last night this exchange of speech became actual. In response to certain signals I admitted to the house a messenger from those outside--a fellow-human, let me hasten to say. He told me much that neither you nor I had even begun to guess, and showed clearly how totally we had misjudged and misinterpreted the purpose of the Outer Ones in maintaining their secret colony on this planet. It seems that the evil legends about what they have offered to men, and what they wish in connection with the earth, are wholly the result of an ignorant misconception of allegorical speech--speech, of course, moulded by cultural backgrounds and thought-habits vastly different from anything we dream of. My own conjectures, I freely own, shot as widely past the mark as any of the guesses of illiterate farmers and savage Indians. What I had thought morbid and shameful and ignominious is in reality awesome and mind-expanding and even glorious--my previous estimate being merely a phase of man's eternal tendency to hate and fear and shrink from the utterly different. Now I regret the harm I have inflicted upon these alien and incredible beings in the course of our nightly skirmishes. If only I had consented to talk peacefully and reasonably with them in the first place! But they bear me no grudge, their emotions being organised very differently from ours. It is their misfortune to have had as their human agents in Vermont some very inferior specimens--the late Walter Brown, for example. He prejudiced me vastly against them. Actually, they have never knowingly harmed men, but have often been cruelly wronged and spied upon by our species. There is a whole secret cult of evil men (a man of your mystical erudition will understand me when I link them with Hastur and the Yellow Sign) devoted to the purpose of tracking them down and injuring them on behalf of monstrous powers from other dimensions. It is against these aggressors--not against normal humanity--that the drastic precautions of the Outer Ones are directed. Incidentally, I learned that many of our lost letters were stolen not by the Outer Ones but by the emissaries of this malign cult. All that the Outer Ones wish of man is peace and non-molestation and an increasing intellectual rapport. This latter is absolutely necessary now that our inventions and devices are expanding our knowledge and motions, and making it more and more impossible for the Outer Ones' necessary outposts to exist secretly on this planet. The alien beings desire to know mankind more fully, and to have a few of mankind's philosophic and scientific leaders know more about them. With such an exchange of knowledge all perils will pass, and a satisfactory modus vivendi be established. The very idea of any attempt to enslave or degrade mankind is ridiculous. As a beginning of this improved rapport, the Outer Ones have naturally chosen me--whose knowledge of them is already so considerable--as their primary interpreter on earth. Much was told me last night--facts of the most stupendous and vista-opening nature--and more will be subsequently communicated to me both orally and in writing. I shall not be called upon to make any trip outside just yet, though I shall probably wish to do so later on--employing special means and transcending everything which we have hitherto been accustomed to regard as human experience. My house will be besieged no longer. Everything has reverted to normal, and the dogs will have no further occupation. In place of terror I have been given a rich boon of knowledge and intellectual adventure which few other mortals have ever shared. The Outer Beings are perhaps the most marvellous organic things in or beyond all space and time-members of a cosmos-wide race of which all other life-forms are merely degenerate variants. They are more vegetable than animal, if these terms can be applied to the sort of matter composing them, and have a somewhat fungoid structure; though the presence of a chlorophyll-like substance and a very singular nutritive system differentiate them altogether from true cormophytic fungi. Indeed, the type is composed of a form of matter totally alien to our part of space--with electrons having a wholly different vibration-rate. That is why the beings cannot be photographed on the ordinary camera films and plates of our known universe, even though our eyes can see them. With proper knowledge, however, any good chemist could make a photographic emulsion which would record their images. The genus is unique in its ability to traverse the heatless and airless interstellar void in full corporeal form, and some of its variants cannot do this without mechanical aid or curious surgical transpositions. Only a few species have the ether-resisting wings characteristic of the Vermont variety. Those inhabiting certain remote peaks in the Old World were brought in other ways. Their external resemblance to animal life, and to the sort of structure we understand as material, is a matter of parallel evolution rather than of close kinship. Their brain-capacity exceeds that of any other surviving life-form, although the winged types of our hill country are by no means the most highly developed. Telepathy is their usual means of discourse, though we have rudimentary vocal organs which, after a slight operation (for surgery is an incredibly expert and everyday thing among them), can roughly duplicate the speech of such types of organism as still use speech. Their main immediate abode is a still undiscovered and almost lightless planet at the very edge of our solar system--beyond Neptune, and the ninth in distance from the sun. It is, as we have inferred, the object mystically hinted at as "Yuggoth" in certain ancient and forbidden writings; and it will soon be the scene of a strange focussing of thought upon our world in an effort to facilitate mental rapport. I would not be surprised if astronomers become sufficiently sensitive to these thought-currents to discover Yuggoth when the Outer Ones wish them to do so. But Yuggoth, of course, is only the stepping-stone. The main body of the beings inhabits strangely organized abysses wholly beyond the utmost reach of any human imagination. The space-time globule which we recognize as the totality of all cosmic entity is only an atom in the genuine infinity which is theirs. And as much of this infinity as any human brain can hold is eventually to be opened up to me, as it has been to not more than fifty other men since the human race has existed. You will probably call this raving at first, Wilmarth, but in time you will appreciate the titanic opportunity I have stumbled upon. I want you to share as much of it as is possible, and to that end must tell you thousands of things that won't go on paper. In the past I have warned you not to come to see me. Now that all is safe, I take pleasure in rescinding that warning and inviting you. Can't you make a trip up here before your college term opens? It would be marvelously delightful if you could. Bring along the phonograph record and all my letters to you as consultative data--we shall need them in piecing together the whole tremendous story. You might bring the Kodak prints, too, since I seem to have mislaid the negatives and my own prints in all this recent excitement. But what a wealth of facts I have to add to all this groping and tentative material--and what a stupendous device I have to supplement my additions! Don't hesitate--I am free from espionage now, and you will not meet anything unnatural or disturbing. Just come along and let my car meet you at the Brattleboro station--prepare to stay as long as you can, and expect many an evening of discussion of things beyond all human conjecture. Don't tell anyone about it, of course--for this matter must not get to the promiscuous public. The train service to Brattleboro is not bad--you can get a timetable in Boston. Take the B. & M. to Greenfield, and then change for the brief remainder of the way. I suggest your taking the convenient 4:10 P.M.--standard-from Boston. This gets into Greenfield at 7:35, and at 9:19 a train leaves there which reaches Brattleboro at 10:01. That is weekdays. Let me know the date and I'll have my car on hand at the station. Pardon this typed letter, but my handwriting has grown shaky of late, as you know, and I don't feel equal to long stretches of script. I got this new Corona in Brattleboro yesterday--it seems to work very well. Awaiting word, and hoping to see you shortly with the

Chapter 5   37

phonograph record and all my letters--and the Kodak prints--I am

Yours in anticipation,

Henry W. Akeley

TO ALBERT N. WILMARTH, ESQ.,

MISKATONIC UNIVERSITY,

ARKHAM, MASS.

The complexity of my emotions upon reading, re-reading, and pondering over this strange and unlooked-for letter is past adequate description. I have said that I was at once relieved and made uneasy, but this expresses only crudely the overtones of diverse and largely subconscious feelings which comprised both the relief and the uneasiness. To begin with, the thing was so antipodally at variance with the whole chain of horrors preceding it--the change of mood from stark terror to cool complacency and even exultation was so unheralded, lightning-like, and complete! I could scarcely believe that a single day could so alter the psychological perspective of one who had written that final frenzied bulletin of Wednesday, no matter what relieving disclosures that day might have brought. At certain moments a sense of conflicting unrealities made me wonder whether this whole distantly reported drama of fantastic forces were not a kind of half-illusory dream created largely within my own mind. Then I thought of the phonograph record and gave way to still greater bewilderment.

The letter seemed so unlike anything which could have been expected! As I analysed my impression, I saw that it consisted of two distinct phases. First, granting that Akeley had been sane before and was still sane, the indicated change in the situation itself was so swift and unthinkable. And secondly, the change in Akeley's own manner, attitude, and language was so vastly beyond the normal or the predictable. The man's whole personality seemed to have undergone an insidious mutation--a mutation so deep that one could scarcely reconcile his two aspects with the supposition that both represented equal sanity. Word-choice, spelling--all were subtly different. And with my academic sensitiveness to prose style, I could trace profound divergences in his commonest reactions and rhythm-responses. Certainly, the emotional cataclysm or revelation which could produce so radical an overturn must be an extreme one indeed! Yet in another way the letter seemed quite characteristic of Akeley. The same old passion for infinity--the same old scholarly inquisitiveness. I could not a moment--or more than a moment--credit the idea of spuriousness or malign substitution. Did not the invitation--the willingness to have me test the truth of the letter in person--prove its genuineness?

I did not retire Saturday night, but sat up thinking of the shadows and marvels behind the letter I had received. My mind, aching from the quick succession of monstrous conceptions it had been forced to confront during the last four months, worked upon this startling new material in a cycle of doubt and acceptance which repeated most of the steps experienced in facing the earlier wonders; till long before dawn a burning interest and curiosity had begun to replace the original storm of perplexity and uneasiness. Mad or sane, metamorphosed or merely relieved, the chances were that Akeley had actually encountered some stupendous change of perspective in his hazardous research; some change at once diminishing his danger--real or fancied--and opening dizzy new vistas of cosmic and superhuman knowledge. My own zeal for the unknown flared up to meet his, and I felt myself touched by the contagion of the morbid barrier-breaking. To shake off the maddening and wearying limitations of time and space and natural law--to be linked with the vast outside--to come close to the nighted and abysmal secrets of the infinite and the ultimate--surely such a thing was worth the risk of one's life, soul, and sanity! And Akeley had said there was no longer any peril--he had invited me to visit him instead of warning me away as before. I tingled at the thought of what he might now have to tell me--there was an almost paralysing fascination in the thought of sitting in that lonely and lately-beleaguered farmhouse with a man who had talked with actual emissaries from outer space; sitting there with the terrible record and the pile of letters in which Akeley had summarised his earlier conclusions.

So late Sunday morning I telegraphed Akeley that I would meet him in Brattleboro on the following Wednesday--September 12th--if that date were convenient for him. In only one respect did I depart from his suggestions, and that concerned the choice of a train. Frankly, I did not feel like arriving in that haunted Vermont region late at night; so instead of accepting the train he chose I telephoned the station and devised another arrangement. By rising early and taking the 8:07 A.M. (standard) into Boston, I could catch the 9:25 for Greenfield; arriving there at 12:22 noon. This connected exactly with a train reaching Brattleboro at 1:08 p.m.--a much more comfortable hour than 10:01 for meeting Akeley and riding with him into the close-packed, secret-guarding hills.

I mentioned this choice in my telegram, and was glad to learn in the reply which came toward evening that it had met with my prospective host's endorsement. His wire ran thus:

ARRANGEMENT SATISFACTORY WILL MEET ONE EIGHT TRAIN WEDNESDAY DONT FORGET RECORD AND LETTERS AND PRINTS KEEP DESTINATION QUIET EXPECT GREAT REVELATIONS AKELEY

Receipt of this message in direct response to one sent to Akeley--and necessarily delivered to his house from the Townshend station either by official messenger or by a restored telephone service--removed any lingering subconscious doubts I may have had about the authorship of the perplexing letter. My relief was marked--indeed, it was greater than I could account for at the time; since all such doubts had been rather deeply buried. But I slept soundly and long that night, and was eagerly busy with preparations during the ensuing two days.

Chapter 6

On Wednesday I started as agreed,. taking with me a valise full of simple necessities and scientific data, including the hideous phonograph record, the Kodak prints, and the entire file of Akeley's correspondence. As requested, I had told no one where I was going; for I could see that the matter demanded utmost privacy, even allowing for its most favourable turns. The thought of actual mental contact with alien, outside entities was stupefying enough to my trained and somewhat prepared mind; and this being so, what might one think of its effect on the vast masses of uninformed laymen? I do not know whether dread or adventurous expectancy was uppermost in me as I changed trains at Boston and began the long westward run out of familiar regions into those I knew less thoroughly. Waltham--Concord--Ayer--Fitchburg--Gardner--Athol -

My train reached Greenfield seven minutes late, but the northbound connecting express had been held. Transferring in haste, I felt a curious breathlessness as the cars rumbled on through the early afternoon sunlight into territories I had always read of but had never before visited. I knew I was entering an altogether older-fashioned and more primitive New England than the mechanised, urbanised coastal and southern areas where all my life had been spent; an unspoiled, ancestral New England without the foreigners and factory-smoke, bill-boards and concrete roads, of the sections which modernity has touched. There would be odd survivals of that continuous native life whose deep roots make it the one authentic outgrowth of the landscape--the continuous native life which keeps alive strange ancient memories, and fertilises the soil for shadowy, marvellous, and seldom-mentioned beliefs.

Now and then I saw the blue Connecticut River gleaming in the sun, and after leaving Northfield we crossed it. Ahead loomed green and cryptical hills, and when the conductor came around I learned that I was at last in Vermont. He told me to set my watch back an hour, since the northern hill country will have no dealings with new-fangled daylight time schemes. As I did so it seemed to me that I was likewise turning the calendar back a century.

Chapter 6

The train kept close to the river, and across in New Hampshire I could see the approaching slope of steep Wantastiquet, about which singular old legends cluster. Then streets appeared on my left, and a green island showed in the stream on my right. People rose and filed to the door, and I followed them. The car stopped, and I alighted beneath the long train-shed of the Brattleboro station.

Looking over the line of waiting motors I hesitated a moment to see which one might turn out to be the Akeley Ford, but my identity was divined before I could take the initiative. And yet it was clearly not Akeley himself who advanced to meet me with an outstretched hand and a mellowly phrased query as to whether I was indeed Mr. Albert N. Wilmarth of Arkham. This man bore no resemblance to the bearded, grizzled Akeley of the snapshot; but was a younger and more urbane person, fashionably dressed, and wearing only a small, dark moustache. His cultivated voice held an odd and almost disturbing hint of vague familiarity, though I could not definitely place it in my memory.

As I surveyed him I heard him explaining that he was a friend of my prospective host's who had come down from Townshend in his stead. Akeley, he declared, had suffered a sudden attack of some asthmatic trouble, and did not feel equal to making a trip in the outdoor air. It was not serious, however, and there was to be no change in plans regarding my visit. I could not make out just how much this Mr. Noyes--as he announced himself--knew of Akeley's researches and discoveries, though it seemed to me that his casual manner stamped him as a comparative outsider. Remembering what a hermit Akeley had been, I was a trifle surprised at the ready availability of such a friend; but did not let my puzzlement deter me from entering the motor to which he gestured me. It was not the small ancient car I had expected from Akeley's descriptions, but a large and immaculate specimen of recent pattern--apparently Noyes's own, and bearing Massachusetts license plates with the amusing "sacred codfish" device of that year. My guide, I concluded, must be a summer transient in the Townshend region.

Noyes climbed into the car beside me and started it at once. I was glad that he did not overflow with conversation, for some peculiar atmospheric tensity made me feel disinclined to talk. The town seemed very attractive in the afternoon sunlight as we swept up an incline and turned to the right into the main street. It drowsed like the older New England cities which one remembers from boyhood, and something in the collocation of roofs and steeples and chimneys and brick walls formed contours touching deep viol-strings of ancestral emotion. I could tell that I was at the gateway of a region half-bewitched through the piling-up of unbroken time-accumulations; a region where old, strange things have had a chance to grow and linger because they have never been stirred up.

As we passed out of Brattleboro my sense of constraint and foreboding increased, for a vague quality in the hill-crowded countryside with its towering, threatening, close-pressing green and granite slopes hinted at obscure secrets and immemorial survivals which might or might not be hostile to mankind. For a time our course followed a broad, shallow river which flowed down from unknown hills in the north, and I shivered when my companion told me it was the West River. It was in this stream, I recalled from newspaper items, that one of the morbid crablike beings had been seen floating after the floods.

Gradually the country around us grew wilder and more deserted. Archaic covered bridges lingered fearsomely out of the past in pockets of the hills, and the half-abandoned railway track paralleling the river seemed to exhale a nebulously visible air of desolation. There were awesome sweeps of vivid valley where great cliffs rose, New England's virgin granite showing grey and austere through the verdure that scaled the crests. There were gorges where untamed streams leaped, bearing down toward the river the unimagined secrets of a thousand pathless peaks. Branching away now and then were narrow, half-concealed roads that bored their way through solid, luxuriant masses of forest among whose primal trees whole armies of elemental spirits might well lurk. As I saw these I thought of how Akeley had been molested by unseen agencies on his drives along this very route, and did not wonder that such things could be.

The quaint, sightly village of Newfane, reached in less than an hour, was our last link with that world which man can definitely call his own by virtue of conquest and complete occupancy. After that we cast off all allegiance to immediate, tangible, and time-touched things, and entered a fantastic world of hushed unreality in which the narrow, ribbon-like road rose and fell and curved with an almost sentient and purposeful caprice amidst the tenantless green peaks and half-deserted valleys. Except for the sound of the motor, and the faint stir of the few lonely farms we passed at infrequent intervals, the only thing that reached my ears was the gurgling, insidious trickle of strange waters from numberless hidden fountains in the shadowy woods.

The nearness and intimacy of the dwarfed, domed hills now became veritably breath-taking. Their steepness and abruptness were even greater than I had imagined from hearsay, and suggested nothing in common with the prosaic objective world we know. The dense, unvisited woods on those inaccessible slopes seemed to harbour alien and incredible things, and I felt that the very outline of the hills themselves held some strange and aeon-forgotten meaning, as if they were vast hieroglyphs left by a rumoured titan race whose glories live only in rare, deep dreams. All the legends of the past, and all the stupefying imputations of Henry Akeley's letters and exhibits, welled up in my memory to heighten the atmosphere of tension and growing menace. The purpose of my visit, and the frightful abnormalities it postulated struck at me all at once with a chill sensation that nearly over-balanced my ardour for strange delvings.

My guide must have noticed my disturbed attitude; for as the road grew wilder and more irregular, and our motion slower and more jolting, his occasional pleasant comments expanded into a steadier flow of discourse. He spoke of the beauty and weirdness of the country, and revealed some acquaintance with the folklore studies of my prospective host. From his polite questions it was obvious that he knew I had come for a scientific purpose, and that I was bringing data of some importance; but he gave no sign of appreciating the depth and awfulness of the knowledge which Akeley had finally reached.

His manner was so cheerful, normal, and urbane that his remarks ought to have calmed and reassured me; but oddly enough. I felt only the more disturbed as we bumped and veered onward into the unknown wilderness of hills and woods. At times it seemed as if he were pumping me to see what I knew of the monstrous secrets of the place, and with every fresh utterance that vague, teasing, baffling familiarity in his voice increased. It was not an ordinary or healthy familiarity despite the thoroughly wholesome and cultivated nature of the voice. I somehow linked it with forgotten nightmares, and felt that I might go mad if I recognised it. If any good excuse had existed, I think I would have turned back from my visit. As it was, I could not well do so--and it occurred to me that a cool, scientific conversation with Akeley himself after my arrival would help greatly to pull me together.

Besides, there was a strangely calming element of cosmic beauty in the hypnotic landscape through which we climbed and plunged fantastically. Time had lost itself in the labyrinths behind, and around us stretched only the flowering waves of faery and the recaptured loveliness of vanished centuries--the hoary groves, the untainted pastures edged with gay autumnal blossoms, and at vast intervals the small brown farmsteads nestling amidst huge trees beneath vertical precipices of fragrant brier and meadow-grass. Even the sunlight assumed a supernal glamour, as if some special atmosphere or exhalation mantled the whole region. I had seen nothing like it before save in the magic vistas that sometimes form the backgrounds of Italian primitives. Sodoma and Leonardo conceived such expanses, but only in the distance, and through the vaultings of Renaissance arcades. We were now burrowing bodily through the midst of the picture, and I seemed to find in its necromancy a thing I had innately known or inherited and for which I had always been vainly searching.

Suddenly, after rounding an obtuse angle at the top of a sharp ascent, the car came to a standstill. On my left, across a well-kept lawn which stretched to the road and flaunted a border of whitewashed stones, rose a white, two-and-a-half-story house of unusual size and elegance for the region, with a congenes of contiguous or arcade-linked barns, sheds, and windmill behind and to the right. I recognised it at once from the snapshot I had received, and was not surprised to see the name of Henry Akeley on the galvanised-iron mailbox near the road. For some distance back of the house a level stretch of marshy and sparsely-wooded land extended, beyond which soared a steep, thickly-forested hillside ending in a jagged leafy crest. This latter, I knew, was the summit of Dark Mountain, half way up which we must have climbed already.

Chapter 6

Alighting from the car and taking my valise, Noyes asked me to wait while he went in and notified Akeley of my advent. He himself, he added, had important business elsewhere, and could not stop for more than a moment. As he briskly walked up the path to the house I climbed out of the car myself, wishing to stretch my legs a little before settling down to a sedentary conversation. My feeling of nervousness and tension had risen to a maximum again now that I was on the actual scene of the morbid beleaguering described so hauntingly in Akeley's letters, and I honestly dreaded the coming discussions which were to link me with such alien and forbidden worlds.

Close contact with the utterly bizarre is often more terrifying than inspiring, and it did not cheer me to think that this very bit of dusty road was the place where those monstrous tracks and that foetid green ichor had been found after moonless nights of fear and death. Idly I noticed that none of Akeley's dogs seemed to be about. Had he sold them all as soon as the Outer Ones made peace with him? Try as I might, I could not have the same confidence in the depth and sincerity of that peace which appeared in Akeley's final and queerly different letter. After all, he was a man of much simplicity and with little worldly experience. Was there not, perhaps, some deep and sinister undercurrent beneath the surface of the new alliance?

Led by my thoughts, my eyes turned downward to the powdery road surface which had held such hideous testimonies. The last few days had been dry, and tracks of all sorts cluttered the rutted, irregular highway despite the unfrequented nature of the district. With a vague curiosity I began to trace the outline of some of the heterogeneous impressions, trying meanwhile to curb the flights of macabre fancy which the place and its memories suggested. There was something menacing and uncomfortable in the funereal stillness, in the muffled, subtle trickle of distant brooks, and in the crowding green peaks and black-wooded precipices that choked the narrow horizon.

And then an image shot into my consciousness which made those vague menaces and flights of fancy seem mild and insignificant indeed. I have said that I was scanning the miscellaneous prints in the road with a kind of idle curiosity--but all at once that curiosity was shockingly snuffed out by a sudden and paralysing gust of active terror. For though the dust tracks were in general confused and overlapping, and unlikely to arrest any casual gaze, my restless vision had caught certain details near the spot where the path to the house joined the highway; and had recognised beyond doubt or hope the frightful significance of those details. It was not for nothing, alas, that I had pored for hours over the Kodak views of the Outer Ones' claw-prints which Akeley had sent. Too well did I know the marks of those loathsome nippers, and that hint of ambiguous direction which stamped the horrors as no creatures of this planet. No chance had been left me for merciful mistake. Here, indeed, in objective form before my own eyes, and surely made not many hours ago, were at least three marks which stood out blasphemously among the surprising plethora of blurred footprints leading to and from the Akeley farmhouse. They were the hellish tracks of the living fungi from Yuggoth.

I pulled myself together in time to stifle a scream. After all, what more was there than I might have expected, assuming that I had really believed Akeley's letters? He had spoken of making peace with the things. Why, then, was it strange that some of them had visited his house? But the terror was stronger than the reassurance. Could any man be expected to look unmoved for the first time upon the claw-marks of animate beings from outer depths of space? Just then I saw Noyes emerge from the door and approach with a brisk step. I must, I reflected, keep command of myself, for the chances were that this genial friend knew nothing of Akeley's profoundest and most stupendous probings into the forbidden.

Akeley, Noyes hastened to inform me, was glad and ready to see me; although his sudden attack of asthma would prevent him from being a very competent host for a day or two. These spells hit him hard when they came, and were always accompanied by a debilitating fever and general weakness. He never was good for much while they lasted--had to talk in a whisper, and was very clumsy and feeble in getting about. His feet and ankles swelled, too, so that he had to bandage them like a gouty old beef-eater. Today he was in rather bad shape, so that I would have to attend very largely to my own needs; but he was none the less eager for conversation. I would find him in the study at the left of the front hall--the room where the blinds were shut. He had to keep

Chapter 6

the sunlight out when he was ill, for his eyes were very sensitive.

As Noyes bade me adieu and rode off northward in his car I began to walk slowly toward the house. The door had been left ajar for me; but before approaching and entering I cast a searching glance around the whole place, trying to decide what had struck me as so intangibly queer about it. The barns and sheds looked trimly prosaic enough, and I noticed Akeley's battered Ford in its capacious, unguarded shelter. Then the secret of the queerness reached me. It was the total silence. Ordinarily a farm is at least moderately murmurous from its various kinds of livestock, but here all signs of life were missing. What of the hens and the dogs? The cows, of which Akeley had said he possessed several, might conceivably be out to pasture, and the dogs might possibly have been sold; but the absence of any trace of cackling or grunting was truly singular.

I did not pause long on the path, but resolutely entered the open house door and closed it behind me. It had cost me a distinct psychological effort to do so, and now that I was shut inside I had a momentary longing for precipitate retreat. Not that the place was in the least sinister in visual suggestion; on the contrary, I thought the graceful late-colonial hallway very tasteful and wholesome, and admired the evident breeding of the man who had furnished it. What made me wish to flee was something very attenuated and indefinable. Perhaps it was a certain odd odour which I thought I noticed--though I well knew how common musty odours are in even the best of ancient farmhouses.

Chapter 7

Refusing to let these cloudy qualms overmaster me, I recalled Noyes's instructions and pushed open the six-panelled, brass-latched white door on my left. The room beyond was darkened as I had known before; and as I entered it I noticed that the queer odour was stronger there. There likewise appeared to be some faint, half-imaginary rhythm or vibration in the air. For a moment the closed blinds allowed me to see very little, but then a kind of apologetic hacking or whispering sound drew my attention to a great easy-chair in the farther, darker corner of the room. Within its shadowy depths I saw the white blur of a man's face and hands; and in a moment I had crossed to greet the figure who had tried to speak. Dim though the light was, I perceived that this was indeed my host. I had studied the Kodak picture repeatedly, and there could be no mistake about this firm, weather-beaten face with the cropped, grizzled beard.

But as I looked again my recognition was mixed with sadness and anxiety; for certainly, his face was that of a very sick man. I felt that there must be something more than asthma behind that strained, rigid, immobile expression and unwinking glassy stare; and realised how terribly the strain of his frightful experiences must have told on him. Was it not enough to break any human being--even a younger man than this intrepid delver into the forbidden? The strange and sudden relief, I feared, had come too late to save him from something like a general breakdown. There was a touch of the pitiful in the limp, lifeless way his lean hands rested in his lap. He had on a loose dressing-gown, and was swathed around the head and high around the neck with a vivid yellow scarf or hood.

And then I saw that he was trying to talk in the same hacking whisper with which he had greeted me. It was a hard whisper to catch at first, since the grey moustache concealed all movements of the lips, and something in its timbre disturbed me greatly; but by concentrating my attention I could soon make out its purport surprisingly well. The accent was by no means a rustic one, and the language was even more polished than correspondence had led me to expect.

Chapter 7

"Mr. Wilmarth, I presume? You must pardon my not rising. I am quite ill, as Mr. Noyes must have told you; but I could not resist having you come just the same. You know what I wrote in my last letter--there is so much to tell you tomorrow when I shall feel better. I can't say how glad I am to see you in person after all our many letters. You have the file with you, of course? And the Kodak prints and records? Noyes put your valise in the hall--I suppose you saw it. For tonight I fear you'll have to wait on yourself to a great extent. Your room is upstairs--the one over this--and you'll see the bathroom door open at the head of the staircase. There's a meal spread for you in the dining-room--right through this door at your right--which you can take whenever you feel like it. I'll be a better host tomorrow--but just now weakness leaves me helpless.

"Make yourself at home--you might take out the letters and pictures and records and put them on the table here before you go upstairs with your bag. It is here that we shall discuss them--you can see my phonograph on that corner stand.

"No, thanks--there's nothing you can do for me. I know these spells of old. Just come back for a little quiet visiting before night, and then go to bed when you please. I'll rest right here--perhaps sleep here all night as I often do. In the morning I'll be far better able to go into the things we must go into. You realise, of course, the utterly stupendous nature of the matter before us. To us, as to only a few men on this earth, there will be opened up gulfs of time and space and knowledge beyond anything within the conception of human science or philosophy.

"Do you know that Einstein is wrong, and that certain objects and forces can move with a velocity greater than that of light? With proper aid I expect to go backward and forward in time, and actually see and feel the earth of remote past and future epochs. You can't imagine the degree to which those beings have carried science. There is nothing they can't do with the mind and body of living organisms. I expect to visit other planets, and even other stars and galaxies. The first trip will be to Yuggoth, the nearest world fully peopled by the beings. It is a strange dark orb at the very rim of our solar system--unknown to earthly astronomers as yet. But I must have written you about this. At

Chapter 7 50

the proper time, you know, the beings there will direct thought-currents toward us and cause it to be discovered--or perhaps let one of their human allies give the scientists a hint.

"There are mighty cities on Yuggoth--great tiers of terraced towers built of black stone like the specimen I tried to send you. That came from Yuggoth. The sun shines there no brighter than a star, but the beings need no light. They have other subtler senses, and put no windows in their great houses and temples. Light even hurts and hampers and confuses them, for it does not exist at all in the black cosmos outside time and space where they came from originally. To visit Yuggoth would drive any weak man mad--yet I am going there. The black rivers of pitch that flow under those mysterious cyclopean bridges--things built by some elder race extinct and forgotten before the beings came to Yuggoth from the ultimate voids--ought to be enough to make any man a Dante or Poe if he can keep sane long enough to tell what he has seen.

"But remember--that dark world of fungoid gardens and windowless cities isn't really terrible. It is only to us that it would seem so. Probably this world seemed just as terrible to the beings when they first explored it in the primal age. You know they were here long before the fabulous epoch of Cthulhu was over, and remember all about sunken R'lyeh when it was above the waters. They've been inside the earth, too--there are openings which human beings know nothing of--some of them in these very Vermont hills--and great worlds of unknown life down there; blue-litten K'n-yan, red-litten Yoth, and black, lightless N'kai. It's from N'kai that frightful Tsathoggua came--you know, the amorphous, toad-like god-creature mentioned in the Pnakotic Manuscripts and the Necronomicon and the Commoriom myth-cycle preserved by the Atlantean high-priest Klarkash-Ton.

"But we will talk of all this later on. It must be four or five o'clock by this time. Better bring the stuff from your bag, take a bite, and then come back for a comfortable chat."

Very slowly I turned and began to obey my host; fetching my valise, extracting and depositing the desired articles, and finally ascending to the room designated as mine. With the memory of that roadside claw-print fresh in my mind, Akeley's whispered paragraphs had affected me queerly; and the hints of familiarity with this unknown world of fungous life--forbidden Yuggoth--made my flesh creep more than I cared to own. I was tremendously sorry about Akeley's illness, but had to confess that his hoarse whisper had a hateful as well as pitiful quality. If only he wouldn't gloat so about Yuggoth and its black secrets!

My room proved a very pleasant and well-furnished one, devoid alike of the musty odour and disturbing sense of vibration; and after leaving my valise there I descended again to greet Akeley and take the lunch he had set out for me. The dining-room was just beyond the study, and I saw that a kitchen ell extended still farther in the same direction. On the dining-table an ample array of sandwiches, cake, and cheese awaited me, and a Thermos-bottle beside a cup and saucer testified that hot coffee had not been forgotten. After a well-relished meal I poured myself a liberal cup of coffee, but found that the culinary standard had suffered a lapse in this one detail. My first spoonful revealed a faintly unpleasant acrid taste, so that I did not take more. Throughout the lunch I thought of Akeley sitting silently in the great chair in the darkened next room.

Once I went in to beg him to share the repast, but he whispered that he could eat nothing as yet. Later on, just before he slept, he would take some malted milk--all he ought to have that day.

After lunch I insisted on clearing the dishes away and washing them in the kitchen sink--incidentally emptying the coffee which I had not been able to appreciate. Then returning to the darkened study I drew up a chair near my host's corner and prepared for such conversation as he might feel inclined to conduct. The letters, pictures, and record were still on the large centre-table, but for the nonce we did not have to draw upon them. Before long I forgot even the bizarre odour and curious suggestions of vibration.

I have said that there were things in some of Akeley's letters--especially the second and most voluminous one--which I would not dare to quote or even form into words on paper. This hesitancy applies with still greater force to the things I heard whispered that evening in the darkened room among the lonely hills. Of the extent of the cosmic horrors unfolded by that raucous voice I cannot even hint. He had known hideous things before, but what he had learned since making his pact with the Outside Things was almost too much for sanity to bear. Even now I absolutely refused to believe what he implied about the constitution of ultimate infinity, the juxtaposition of dimensions, and the frightful position of our known cosmos of space and time in the unending chain of linked cosmos-atoms which makes up the immediate super-cosmos of curves, angles, and material and semi-material electronic organisation.

Never was a sane man more dangerously close to the arcana of basic entity--never was an organic brain nearer to utter annihilation in the chaos that transcends form and force and symmetry. I learned whence Cthulhu first came, and why half the great temporary stars of history had flared forth. I guessed--from hints which made even my informant pause timidly--the secret behind the Magellanic Clouds and globular nebulae, and the black truth veiled by the immemorial allegory of Tao. The nature of the Doels was plainly revealed, and I was told the essence (though not the source) of the Hounds of Tindalos. The legend of Yig, Father of Serpents, remained figurative no longer, and I started with loathing when told of the monstrous nuclear chaos beyond angled space which the Necronomicon had mercifully cloaked under the name of Azathoth. It was shocking to have the foulest nightmares of secret myth cleared up in concrete terms whose stark, morbid hatefulness exceeded the boldest hints of ancient and mediaeval mystics. Ineluctably I was led to believe that the first whisperers of these accursed tales must have had discourse with Akeley's Outer Ones, and perhaps have visited outer cosmic realms as Akeley now proposed visiting them.

I was told of the Black Stone and what it implied, and was glad that it had not reached me. My guesses about those hieroglyphics had been all too correct! And yet Akeley now seemed reconciled to the whole fiendish system he had stumbled upon; reconciled and eager to probe farther into the monstrous abyss. I wondered what beings he had talked with since his last letter to me, and whether many of them had been as human as that first emissary he had mentioned. The tension in my head grew insufferable, and I built up all sorts of wild theories

Chapter 7

about that queer, persistent odour and those insidious hints of vibration in the darkened room.

Night was falling now, and as I recalled what Akeley had written me about those earlier nights I shuddered to think there would be no moon. Nor did I like the way the farmhouse nestled in the lee of that colossal forested slope leading up to Dark Mountain's unvisited crest. With Akeley's permission I lighted a small oil lamp, turned it low, and set it on a distant bookcase beside the ghostly bust of Milton; but afterward I was sorry I had done so, for it made my host's strained, immobile face and listless hands look damnably abnormal and corpselike. He seemed half-incapable of motion, though I saw him nod stiffly once in awhile.

After what he had told, I could scarcely imagine what profounder secrets he was saving for the morrow; but at last it developed that his trip to Yuggoth and beyond--and my own possible participation in it--was to be the next day's topic. He must have been amused by the start of horror I gave at hearing a cosmic voyage on my part proposed, for his head wabbled violently when I showed my fear. Subsequently he spoke very gently of how human beings might accomplish--and several times had accomplished--the seemingly impossible flight across the interstellar void. It seemed that complete human bodies did not indeed make the trip, but that the prodigious surgical, biological, chemical, and mechanical skill of the Outer Ones had found a way to convey human brains without their concomitant physical structure.

There was a harmless way to extract a brain, and a way to keep the organic residue alive during its absence. The bare, compact cerebral matter was then immersed in an occasionally replenished fluid within an ether-tight cylinder of a metal mined in Yuggoth, certain electrodes reaching through and connecting at will with elaborate instruments capable of duplicating the three vital faculties of sight, hearing, and speech. For the winged fungus-beings to carry the brain-cylinders intact through space was an easy matter. Then, on every planet covered by their civilisation, they would find plenty of adjustable faculty-instruments capable of being connected with the encased brains; so that after a little fitting these travelling intelligences could be given a full sensory and articulate life--albeit a bodiless and

Chapter 7

mechanical one--at each stage of their journeying through and beyond the space-time continuum. It was as simple as carrying a phonograph record about and playing it wherever a phonograph of corresponding make exists. Of its success there could be no question. Akeley was not afraid. Had it not been brilliantly accomplished again and again?

For the first time one of the inert, wasted hands raised itself and pointed stiffly to a high shelf on the farther side of the room. There, in a neat row, stood more than a dozen cylinders of a metal I had never seen before--cylinders about a foot high and somewhat less in diameter, with three curious sockets set in an isosceles triangle over the front convex surface of each. One of them was linked at two of the sockets to a pair of singular-looking machines that stood in the background. Of their purport I did not need to be told, and I shivered as with ague. Then I saw the hand point to a much nearer corner where some intricate instruments with attached cords and plugs, several of them much like the two devices on the shelf behind the cylinders, were huddled together.

"There are four kinds of instruments here, Wilmarth," whispered the voice. "Four kinds--three faculties each--makes twelve pieces in all. You see there are four different sorts of beings represented in those cylinders up there. Three humans, six fungoid beings who can't navigate space corporeally, two beings from Neptune (God! if you could see the body this type has on its own planet!), and the rest entities from the central caverns of an especially interesting dark star beyond the galaxy. In the principal outpost inside Round Hill you'll now and then find more cylinders and machines--cylinders of extra-cosmic brains with different senses from any we know--allies and explorers from the uttermost Outside--and special machines for giving them impressions and expression in the several ways suited at once to them and to the comprehensions of different types of listeners. Round Hill, like most of the beings' main outposts all through the various universes, is a very cosmopolitan place. Of course, only the more common types have been lent to me for experiment.

"Here--take the three machines I point to and set them on the table. That tall one with the two glass lenses in front--then the box with the vacuum tubes and sounding-board--and now the one with the metal

Chapter 7   55

disc on top. Now for the cylinder with the label 'B-67' pasted on it. Just stand in that Windsor chair to reach the shelf. Heavy? Never mind! Be sure of the number--B-67. Don't bother that fresh, shiny cylinder joined to the two testing instruments--the one with my name on it. Set B-67 on the table near where you've put the machines--and see that the dial switch on all three machines is jammed over to the extreme left.

"Now connect the cord of the lens machine with the upper socket on the cylinder--there! Join the tube machine to the lower left-hand socket, and the disc apparatus to the outer socket. Now move all the dial switches on the machine over to the extreme right--first the lens one, then the disc one, and then the tube one. That's right. I might as well tell you that this is a human being--just like any of us. I'll give you a taste of some of the others tomorrow."

To this day I do not know why I obeyed those whispers so slavishly, or whether I thought Akeley was mad or sane. After what had gone before, I ought to have been prepared for anything; but this mechanical mummery seemed so like the typical vagaries of crazed inventors and scientists that it struck a chord of doubt which even the preceding discourse had not excited. What the whisperer implied was beyond all human belief--yet were not the other things still farther beyond, and less preposterous only because of their remoteness from tangible concrete proof?

As my mind reeled amidst this chaos, I became conscious of a mixed grating and whirring from all three of the machines lately linked to the cylinder--a grating and whirring which soon subsided into a virtual noiselessness. What was about to happen? Was I to hear a voice? And if so, what proof would I have that it was not some cleverly concocted radio device talked into by a concealed but closely watched speaker? Even now I am unwilling to swear just what I heard, or just what phenomenon really took place before me. But something certainly seemed to take place.

To be brief and plain, the machine with the tubes and sound-box began to speak, and with a point and intelligence which left no doubt that the speaker was actually present and observing us. The voice was loud, metallic, lifeless, and plainly mechanical in every detail of its

Chapter 7

production. It was incapable of inflection or expressiveness, but scraped and rattled on with a deadly precision and deliberation.

"Mr. Wilmarth," it said, "I hope I do not startle you. I am a human being like yourself, though my body is now resting safely under proper vitalising treatment inside Round Hill, about a mile and a half east of here. I myself am here with you--my brain is in that cylinder and I see, hear, and speak through these electronic vibrators. In a week I am going across the void as I have been many times before, and I expect to have the pleasure of Mr. Akeley's company. I wish I might have yours as well; for I know you by sight and reputation, and have kept close track of your correspondence with our friend. I am, of course, one of the men who have become allied with the outside beings visiting our planet. I met them first in the Himalayas, and have helped them in various ways. In return they have given me experiences such as few men have ever had.

"Do you realise what it means when I say I have been on thirty-seven different celestial bodies--planets, dark stars, and less definable objects--including eight outside our galaxy and two outside the curved cosmos of space and time? All this has not harmed me in the least. My brain has been removed from my body by fissions so adroit that it would be crude to call the operation surgery. The visiting beings have methods which make these extractions easy and almost normal--and one's body never ages when the brain is out of it. The brain, I may add, is virtually immortal with its mechanical faculties and a limited nourishment supplied by occasional changes of the preserving fluid.

"Altogether, I hope most heartily that you will decide to come with Mr. Akeley and me. The visitors are eager to know men of knowledge like yourself, and to show them the great abysses that most of us have had to dream about in fanciful ignorance. It may seem strange at first to meet them, but I know you will be above minding that. I think Mr. Noyes will go along, too--the man who doubtless brought you up here in his car. He has been one of us for years--I suppose you recognised his voice as one of those on the record Mr. Akeley sent you."

At my violent start the speaker paused a moment before concluding. "So Mr. Wilmarth, I will leave the matter to you; merely adding that a

Chapter 7

man with your love of strangeness and folklore ought never to miss such a chance as this. There is nothing to fear. All transitions are painless; and there is much to enjoy in a wholly mechanised state of sensation. When the electrodes are disconnected, one merely drops off into a sleep of especially vivid and fantastic dreams.

"And now, if you don't mind, we might adjourn our session till tomorrow. Good night--just turn all the switches back to the left; never mind the exact order, though you might let the lens machine be last. Good night, Mr. Akeley--treat our guest well! Ready now with those switches?"

That was all. I obeyed mechanically and shut off all three switches, though dazed with doubt of everything that had occurred. My head was still reeling as I heard Akeley's whispering voice telling me that I might leave all the apparatus on the table just as it was. He did not essay any comment on what had happened, and indeed no comment could have conveyed much to my burdened faculties. I heard him telling me I could take the lamp to use in my room, and deduced that he wished to rest alone in the dark. It was surely time he rested, for his discourse of the afternoon and evening had been such as to exhaust even a vigorous man. Still dazed, I bade my host good night and went upstairs with the lamp, although I had an excellent pocket flashlight with me.

I was glad to be out of that downstairs study with the queer odour and vague suggestions of vibration, yet could not of course escape a hideous sense of dread and peril and cosmic abnormality as I thought of the place I was in and the forces I was meeting. The wild, lonely region, the black, mysteriously forested slope towering so close behind the house; the footprint in the road, the sick, motionless whisperer in the dark, the hellish cylinders and machines, and above all the invitations to strange surgery and stranger voyagings--these things, all so new and in such sudden succession, rushed in on me with a cumulative force which sapped my will and almost undermined my physical strength.

To discover that my guide Noyes was the human celebrant in that monstrous bygone Sabbat-ritual on the phonograph record was a particular shock, though I had previously sensed a dim, repellent familiarity in his voice. Another special shock came from my own attitude toward my host whenever I paused to analyse it; for much as I had instinctively liked Akeley as revealed in his correspondence, I now found that he filled me with a distinct repulsion. His illness ought to have excited my pity; but instead, it gave me a kind of shudder. He was so rigid and inert and corpselike--and that incessant whispering was so hateful and unhuman!

It occurred to me that this whispering was different from anything else of the kind I had ever heard; that, despite the curious motionlessness of the speaker's moustache-screened lips, it had a latent strength and carrying-power remarkable for the wheezing of an asthmatic. I had been able to understand the speaker when wholly across the room, and once or twice it had seemed to me that the faint but penetrant sounds represented not so much weakness as deliberate repression--for what reason I could not guess. From the first I had felt a disturbing quality in their timbre. Now, when I tried to weigh the matter, I thought I could trace this impression to a kind of subconscious familiarity like that which had made Noyes's voice so hazily ominous. But when or where I had encountered the thing it hinted at, was more than I could tell.

One thing was certain--I would not spend another night here. My scientific zeal had vanished amidst fear and loathing, and I felt nothing now but a wish to escape from this net of morbidity and unnatural revelation. I knew enough now. It must indeed be true that strange cosmic linkages do exist--but such things are surely not meant for normal human beings to meddle with.

Blasphemous influences seemed to surround me and press chokingly upon my senses. Sleep, I decided, would be out of the question; so I merely extinguished the lamp and threw myself on the bed fully dressed. No doubt it was absurd, but I kept ready for some unknown emergency; gripping in my right hand the revolver I had brought along, and holding the pocket flashlight in my left. Not a sound came from below, and I could imagine how my host was sitting there with cadaverous stiffness in the dark.

Chapter 7

Somewhere I heard a clock ticking, and was vaguely grateful for the normality of the sound. It reminded me, though, of another thing about the region which disturbed me--the total absence of animal life. There were certainly no farm beasts about, and now I realised that even the accustomed night-noises of wild living things were absent. Except for the sinister trickle of distant unseen waters, that stillness was anomalous--interplanetary--and I wondered what star-spawned, intangible blight could be hanging over the region. I recalled from old legends that dogs and other beasts had always hated the Outer Ones, and thought of what those tracks in the road might mean.

Chapter 8

Do not ask me how long my unexpected lapse into slumber lasted, or how much of what ensued was sheer dream. If I tell you that I awakened at a certain time, and heard and saw certain things, you will merely answer that I did not wake then; and that everything was a dream until the moment when I rushed out of the house, stumbled to the shed where I had seen the old Ford, and seized that ancient vehicle for a mad, aimless race over the haunted hills which at last landed me--after hours of jolting and winding through forest-threatened labyrinths--in a village which turned out to be Townshend.

You will also, of course, discount everything else in my report; and declare that all the pictures, record-sounds, cylinder-and-machine sounds, and kindred evidences were bits of pure deception practiced on me by the missing Henry Akeley. You will even hint that he conspired with other eccentrics to carry out a silly and elaborate hoax--that he had the express shipment removed at Keene, and that he had Noyes make that terrifying wax record. It is odd, though, that Noyes has not ever yet been identified; that he was unknown at any of the villages near Akeley's place, though he must have been frequently in the region. I wish I had stopped to memorize the license-number of his car--or perhaps it is better after all that I did not. For I, despite all you can say, and despite all I sometimes try to say to myself, know that loathsome outside influences must be lurking there in the half-unknown hills--and that, those influences have spies and emissaries in the world of men. To keep as far as possible from such influences and such emissaries is all that I ask of life in future.

When my frantic story sent a sheriff's posse out to the farmhouse, Akeley was gone without leaving a trace. His loose dressing gown, yellow scarf, and foot-bandages lay on the study floor near his corner easy-chair, and it could not be decided whether any of his other apparel had vanished with him. The dogs and livestock were indeed missing, and there were some curious bullet-holes both on the house's exterior and on some of the walls within; but beyond this nothing unusual could be detected. No cylinders or machines, none of the evidences I had brought in my valise, no queer odour or vibration-sense, no foot-prints in the road, and none of the problematical things I glimpsed at the very last.

I stayed a week in Brattleboro after my escape, making inquiries among people of every kind who had known Akeley; and the results convince me that the matter is no figment of dream or delusion. Akeley's queer purchase of dogs and ammunition and chemicals, and the cutting of his telephone wires, are matters of record; while all who knew him--including his son in California--concede that his occasional remarks on strange studies had a certain consistency. Solid citizens believe he was mad, and unhesitatingly pronounce all reported evidences mere hoaxes devised with insane cunning and perhaps abetted by eccentric associates; but the lowlier country folk sustain his statements in every detail. He had showed some of these rustics his photographs and black stone, and had played the hideous record for them; and they all said the footprints and buzzing voice were like those described in ancestral legends.

They said, too, that suspicious sights and sounds had been noticed increasingly around Akeley's house after he found the black stone, and that the place was now avoided by everybody except the mail man and other casual, tough-minded people. Dark Mountain and Round Hill were both notoriously haunted spots, and I could find no one who had ever closely explored either. Occasional disappearances of natives throughout the district's history were well attested, and these now included the semi-vagabond Walter Brown, whom Akeley's letters had mentioned. I even came upon one farmer who thought he had personally glimpsed one of the queer bodies at flood-time in the swollen West River, but his tale was too confused to be really valuable.

When I left Brattleboro I resolved never to go back to Vermont, and I feel quite certain I shall keep my resolution. Those wild hills are surely the outpost of a frightful cosmic race--as I doubt all the less since reading that a new ninth planet has been glimpsed beyond Neptune, just as those influences had said it would be glimpsed. Astronomers, with a hideous appropriateness they little suspect, have named this thing "Pluto." I feel, beyond question, that it is nothing less than nighted Yuggoth--and I shiver when I try to figure out the real reason why its monstrous denizens wish it to be known in this way at this especial time. I vainly try to assure myself that these daemoniac creatures are

Chapter 8

not gradually leading up to some new policy hurtful to the earth and its normal inhabitants.

But I have still to tell of the ending of that terrible night in the farmhouse. As I have said, I did finally drop into a troubled doze; a doze filled with bits of dream which involved monstrous landscape-glimpses. Just what awaked me I cannot yet say, but that I did indeed awake at this given point I feel very certain. My first confused impression was of stealthily creaking floor-boards in the hall outside my door, and of a clumsy, muffled fumbling at the latch. This, however, ceased almost at once; so that my really clear impressions begin with the voices heard from the study below. There seemed to be several speakers, and I judged that they were controversially engaged.

By the time I had listened a few seconds I was broad awake, for the nature of the voices was such as to make all thought of sleep ridiculous. The tones were curiously varied, and no one who had listened to that accursed phonograph record could harbour any doubts about the nature of at least two of them. Hideous though the idea was, I knew that I was under the same roof with nameless things from abysmal space; for those two voices were unmistakably the blasphemous buzzings which the Outside Beings used in their communication with men. The two were individually different--different in pitch, accent, and tempo--but they were both of the same damnable general kind.

A third voice was indubitably that of a mechanical utterance-machine connected with one of the detached brains in the cylinders. There was as little doubt about that as about the buzzings; for the loud, metallic, lifeless voice of the previous evening, with its inflectionless, expressionless scraping and rattling, and its impersonal precision and deliberation, had been utterly unforgettable. For a time I did not pause to question whether the intelligence behind the scraping was the identical one which had formerly talked to me; but shortly afterward I reflected that any brain would emit vocal sounds of the same quality if linked to the same mechanical speech-producer; the only possible differences being in language, rhythm, speed, and pronunciation. To complete the eldritch colloquy there were two actually human voices--one the crude speech of an unknown and evidently rustic man,

Chapter 8

and the other the suave Bostonian tones of my erstwhile guide Noyes.

As I tried to catch the words which the stoutly-fashioned floor so bafflingly intercepted, I was also conscious of a great deal of stirring and scratching and shuffling in the room below; so that I could not escape the impression that it was full of living beings--many more than the few whose speech I could single out. The exact nature of this stirring is extremely hard to describe, for very few good bases of comparison exist. Objects seemed now and then to move across the room like conscious entities; the sound of their footfalls having something about it like a loose, hard-surfaced clattering--as of the contact of ill-coordinated surfaces of horn or hard rubber. It was, to use a more concrete but less accurate comparison, as if people with loose, splintery wooden shoes were shambling and rattling about on the polished board floor. Of the nature and appearance of those responsible for the sounds, I did not care to speculate.

Before long I saw that it would be impossible to distinguish any connected discourse. Isolated words--including the names of Akeley and myself--now and then floated up, especially when uttered by the mechanical speech-producer; but their true significance was lost for want of continuous context. Today I refuse to form any definite deductions from them, and even their frightful effect on me was one of suggestion rather than of revelation. A terrible and abnormal conclave, I felt certain, was assembled below me; but for what shocking deliberations I could not tell. It was curious how this unquestioned sense of the malign and the blasphemous pervaded me despite Akeley's assurances of the Outsider's friendliness.

With patient listening I began to distinguish clearly between voices, even though I could not grasp much of what any of the voices said. I seemed to catch certain typical emotions behind some of the speakers. One of the buzzing voices, for example, held an unmistakable note of authority; whilst the mechanical voice, notwithstanding its artificial loudness and regularity, seemed to be in a position of subordination and pleading. Noyes's tones exuded a kind of conciliatory atmosphere. The others I could make no attempt to interpret. I did not hear the familiar whisper of Akeley, but well knew that such a sound could never penetrate the solid flooring of my room.

I will try to set down some of the few disjointed words and other sounds I caught, labelling the speakers of the words as best I know how. It was from the speech-machine that I first picked up a few recognisable phrases.

(The Speech-Machine)

"...brought it on myself... sent back the letters and the record... end on it... taken in... seeing and hearing... damn you... impersonal force, after all... fresh, shiny cylinder... great God..."

(First Buzzing Voice)

"...time we stopped... small and human... Akeley... brain... saying..."

(Second Buzzing Voice)

"Nyarlathotep... Wilmarth... records and letters... cheap imposture..."

(Noyes)

"...(an unpronounceable word or name, possibly N'gah-Kthun) harmless... peace... couple of weeks... theatrical... told you that before..."

(First Buzzing Voice)

"...no reason... original plan... effects... Noyes can watch Round Hill... fresh cylinder... Noyes's car..."

(Noyes)

"...well... all yours... down here... rest... place..."

(Several Voices at Once in Indistinguishable Speech)

(Many Footsteps, Including the Peculiar Loose Stirring or Clattering)

(A Curious Sort of Flapping Sound)

Chapter 8

(The Sound of an Automobile Starting and Receding)

(Silence)

That is the substance of what my ears brought me as I lay rigid upon that strange upstairs bed in the haunted farmhouse among the daemoniac hills--lay there fully dressed, with a revolver clenched in my right hand and a pocket flashlight gripped in my left. I became, as I have said, broad awake; but a kind of obscure paralysis nevertheless kept me inert till long after the last echoes of the sounds had died away. I heard the wooden, deliberate ticking of the ancient Connecticut clock somewhere far below, and at last made out the irregular snoring of a sleeper. Akeley must have dozed off after the strange session, and I could well believe that he needed to do so.

Just what to think or what to do was more than I could decide. After all, what had I heard beyond things which previous information might have led me to expect? Had I not known that the nameless Outsiders were now freely admitted to the farmhouse? No doubt Akeley had been surprised by an unexpected visit from them. Yet something in that fragmentary discourse had chilled me immeasurably, raised the most grotesque and horrible doubts, and made me wish fervently that I might wake up and prove everything a dream. I think my subconscious mind must have caught something which my consciousness has not yet recognised. But what of Akeley? Was he not my friend, and would he not have protested if any harm were meant me? The peaceful snoring below seemed to cast ridicule on all my suddenly intensified fears.

Was it possible that Akeley had been imposed upon and used as a lure to draw me into the hills with the letters and pictures and phonograph record? Did those beings mean to engulf us both in a common destruction because we had come to know too much? Again I thought of the abruptness and unnaturalness of that change in the situation which must have occurred between Akeley's penultimate and final letters. Something, my instinct told me, was terribly wrong. All was not as it seemed. That acrid coffee which I refused--had there not been an attempt by some hidden, unknown entity to drug it? I must talk to Akeley at once, and restore his sense of proportion. They had hypnotised him with their promises of cosmic revelations, but now he must listen to reason. We. must get out of this before it would be too late. If he lacked the will power to make the break for liberty. I would supply it. Or if I could not persuade him to go, I could at least go myself. Surely he would let me take his Ford and leave it in a garage in Brattleboro. I had noticed it in the shed--the door being left unlocked and open now that peril was deemed past--and I believed there was a good chance of its being ready for instant use. That momentary dislike of Akeley which I had felt during and after the evening's conversation was all gone now. He was in a position much like my own, and we must stick together. Knowing his indisposed condition, I hated to wake him at this juncture, but I knew that I must. I could not stay in this place till morning as matters stood.

At last I felt able to act, and stretched myself vigorously to regain command of my muscles. Arising with a caution more impulsive than deliberate, I found and donned my hat, took my valise, and started downstairs with the flashlight's aid. In my nervousness I kept the revolver clutched in my right hand, being able to take care of both valise and flashlight with my left. Why I exerted these precautions I do not really know, since I was even then on my way to awaken the only other occupant of the house.

As I half-tiptoed down the creaking stairs to the lower hall I could hear the sleeper more plainly, and noticed that he must be in the room on my left--the living-room I had not entered. On my right was the gaping blackness of the study in which I had heard the voices. Pushing open the unlatched door of the living-room I traced a path with the flashlight toward the source of the snoring, and finally turned the beams on the sleeper's face. But in the next second I hastily turned them away and commenced a catlike retreat to the hall, my caution this time springing from reason as well as from instinct. For the sleeper on the couch was not Akeley at all, but my quondam guide Noyes.

Just what the real situation was, I could not guess; but common sense told me that the safest thing was to find out as much as possible before arousing anybody. Regaining the hall, I silently closed and latched the living-room door after me; thereby lessening the chances of awakening Noyes. I now cautiously entered the dark study, where I

Chapter 8

expected to find Akeley, whether asleep or awake, in the great corner chair which was evidently his favorite resting-place. As I advanced, the beams of my flashlight caught the great centre-table, revealing one of the hellish cylinders with sight and hearing machines attached, and with a speech machine standing close by, ready to be connected at any moment. This, I reflected, must be the encased brain I had heard talking during the frightful conference; and for a second I had a perverse impulse to attach the speech machine and see what it would say.

It must, I thought, be conscious of my presence even now; since the sight and hearing attachments could not fail to disclose the rays of my flashlight and the faint creaking of the floor beneath my feet. But in the end I did not dare meddle with the thing. I idly saw that it was the fresh shiny cylinder with Akeley's name on it, which I had noticed on the shelf earlier in the evening and which my host had told me not to bother. Looking back at that moment, I can only regret my timidity and wish that I had boldly caused the apparatus to speak. God knows what mysteries and horrible doubts and questions of identity it might have cleared up! But then, it may be merciful that I let it alone.

From the table I turned my flashlight to the corner where I thought Akeley was, but found to my perplexity that the great easy-chair was empty of any human occupant asleep or awake. From the seat to the floor there trailed voluminously the familiar old dressing-gown, and near it on the floor lay the yellow scarf and the huge foot-bandages I had thought so odd. As I hesitated, striving to conjecture where Akeley might be, and why he had so suddenly discarded his necessary sick-room garments, I observed that the queer odour and sense of vibration were no longer in the room. What had been their cause? Curiously it occurred to me that I had noticed them only in Akeley's vicinity. They had been strongest where he sat, and wholly absent except in the room with him or just outside the doors of that room. I paused, letting the flashlight wander about the dark study and racking my brain for explanations of the turn affairs had taken.

Would to Heaven I had quietly left the place before allowing that light to rest again on the vacant chair. As it turned out, I did not leave quietly; but with a muffled shriek which must have disturbed, though it did not quite awake, the sleeping sentinel across the hall. That shriek, and Noyes's still-unbroken snore, are the last sounds I ever heard in that morbidity-choked farmhouse beneath the black-wooded crest of haunted mountain--that focus of transcosmic horror amidst the lonely green hills and curse-muttering brooks of a spectral rustic land.

It is a wonder that I did not drop flashlight, valise, and revolver in my wild scramble, but somehow I failed to lose any of these. I actually managed to get out of that room and that house without making any further noise, to drag myself and my belongings safely into the old Ford in the shed, and to set that archaic vehicle in motion toward some unknown point of safety in the black, moonless night. The ride that followed was a piece of delirium out of Poe or Rimbaud or the drawings of Dore, but finally I reached Townshend. That is all. If my sanity is still unshaken, I am lucky. Sometimes I fear what the years will bring, especially since that new planet Pluto has been so curiously discovered.

As I have implied, I let my flashlight return to the vacant easy-chair after its circuit of the room; then noticing for the first time the presence of certain objects in the seat, made inconspicuous by the adjacent loose folds of the empty dressing-gown. These are the objects, three in number, which the investigators did not find when they came later on. As I said at the outset, there was nothing of actual visual horror about them. The trouble was in what they led one to infer. Even now I have my moments of half-doubt--moments in which I half-accept the scepticism of those who attribute my whole experience to dream and nerves and delusion.

The three things were damnably clever constructions of their kind, and were furnished with ingenious metallic clamps to attach them to organic developments of which I dare not form any conjecture. I hope--devoutly hope-that they were the waxen products of a master artist, despite what my inmost fears tell me. Great God! That whisperer in darkness with its morbid odour and vibrations! Sorcerer, emissary, changeling, outsider... that hideous repressed buzzing... and all the time in that fresh, shiny cylinder on the shelf... poor devil... "Prodigious surgical, biological, chemical, and mechanical skill...

Chapter 8

For the things in the chair, perfect to the last, subtle detail of microscopic resemblance--or identity--were the face and hands of Henry Wentworth Akeley.

2 RTEXTR*ch

Darkness

A free ebook from http://manybooks.net/

www.ingramcontent.com/pod-product-compliance
Lightning Source LLC
Chambersburg PA
CBHW050240230526
45470CB00005B/2038